VOYAGE THROUGH THE ANTARCTIC

RICHARD ADAMS AND RONALD LOCKLEY

VOYAGE THROUGH

ALFRED A. KNOPF   NEW YORK   1983

THE ANTARCTIC

PHOTOGRAPHS BY PETER HIRST-SMITH

THIS IS A BORZOI BOOK
PUBLISHED BY ALFRED A. KNOPF, INC.
Copyright © 1982 by Richard Adams and Ronald Lockley
Photographs Copyright © 1982 by Peter Hirst-Smith
All rights reserved under International and Pan-American
Copyright Conventions.
Published in the United States by
Alfred A. Knopf, Inc., New York.
Distributed by Random House, Inc., New York.
Published in Great Britain by Allen Lane, Penguin Books, Ltd.

Designed by Paul Bowden
Map drawn by Reginald Piggott
Set in Monophoto Sabon

Library of Congress Cataloging in Publication Data
Adams, Richard, [date]
Voyage through the Antarctic.
1. Antarctic regions—Description and travel.
I. Lockley, R. M. (Ronald Mathias), [date]
II. Title.
G860.A3 1982   919.8′04   82-48484
ISBN 0-394-52858-1
Manufactured in the United States of America
FIRST AMERICAN EDITION

CONTENTS

PREFACE

During our voyage Peter Hirst-Smith took nearly three thousand photographs. To choose fewer than a hundred of these for inclusion in this book was a hard task. We have tried to achieve a balance between landscape (or icescape) and wildlife, but no one feeling disappointed that some creature, plant, scene or natural phenomenon mentioned in the text is not depicted as well, should jump to the conclusion that that is because Peter failed to get a picture. Faced with an *embarras de richesses* and the necessity of producing the book at a reasonable price, we have tried to make the best selection. For the rest, piece out our imperfections with your thoughts.

Passages in italics are by Ronald Lockley.

Richard Adams
Ronald Lockley

160°W

140°W

40°S

50°S

60°S

70°S

Antarctic Circle

180°

Chatham Is

Auckland • Wellington

NEW ZEALAND

ROSS
SEA

Stewart I.   Campbell I.

Snares I.   Ro.
Auckland Is   Ice Sh.
(incl. Enderby I.)   Cape Adare   Cape Hallett

Balleny Is   McMurdo   Scott Bas
Young I.   Sturge Is   Sound   (Cape Eva
                        VICTORIA   Ross I.
160°E   LAND   McMurd

Macquarie I.   Beaufort I. (Hut Po

TASMANIA

AUSTRALIA

140°E

120°E

100°E

0   500   1000 km
0   500 miles

W

100°W

80°W

60°W

Ushuaia

Tierra del Fuego

Cape Horn

Falkland Is

Drake Passage

Peter Ist I.

Amundsen
Sea

Bellingshausen
Sea

Deception I.

Anvers I.

King George I.

Gerlache
Strait

South Orkney Is

40°W

Antarctic
Peninsula

ELLSWORTH
LAND

PALMER LAND

South
Georgia

YRD
AND

Weddell
Sea

Filchner
Ice
Shelf

South Sandwich Is

20°W

South Pole

80°S

D R O N N I N G   M A U D   L A N D

70°S

0°

ENDERBY
LAND

60°S

Antarctic Circle

50°S

Mackenzie
Bay

80°E

60°E

40°E

20°E

1

TRAVELLING OUT

It was a fine, mild day in late January. I was sitting at home in the Isle of Man, contemplating an accumulation of papers – passport, airline tickets, travellers' cheques, mailing instructions, luggage labels, a fifteen-page, illustrated brochure marked *Lindblad Travel Presents Antarctica*, and eleven pages of joining instructions called 'Preparing for Antarctica. Please Read Immediately'. (Does it explode or something if you don't?) Well, here goes: Personal information form – one photo – cholera vaccination (no), yellow fever (no) – tell us your arrival date at Buenos Aires – tell us your departure date from New Zealand. 'Keeping warm and dry in winter weather' by Jane E. Brody. 'Passengers dress informally on board the ship. Jackets and ties are not required. But there will be a few special evenings, both on board and ashore, when the men may *wish* to wear jackets and ties.'

A lot of stuff to assimilate. Daunting or exciting? In my experience, expeditions cease to be daunting and become exciting to the extent that you feel you're properly equipped. Once you've been and gone and got a lot of gear, it's natural to look forward to using it. 'As it is a relatively small vessel, there is no room, other than the passenger cabins, for storing baggage, so please refrain from using oversize suitcases. The chit system is used on board for drinks at the bar.' This sounded all right: the *Shorter Oxford English Dictionary* defines 'chit' as 'a girl or young woman (contemptuous), 1624'.

Obviously, I felt, the proper clothes were all-important for Antarctica; almost more than for Ascot, in some ways. You can't very well slip out and buy something from the penguins if you find you haven't brought it. Surveying the library floor, I reckoned I'd been pretty thorough. Two sets of thermal underwear, two balaclava helmets, two roll-neck sweaters and any number of pullovers. ('For anyone out in the cold, it's far better to wear layers of relatively light, loose clothing than one thick, heavy item. Between each layer there is a film of trapped air. Wool is superior to cotton, because it can trap a lot of air.' Jane E. Brody, no less.)

What I felt proudest of (why? I hadn't made them) were my fluorescent orange, artificial-fur-lined, hooded, double-zip, polyester waterproof outer jacket, big enough to go on over masses of clothing; my padded waterproof trousers ditto; and my moon-boots. The moon-boots had been a tip from Eric Hosking, the famous bird photographer, who had been through the Antarctic before. 'They're not very durable,' said he, 'but they'll certainly last out one trip, and they're very warm and efficient on snow and ice.' Bought, they turned out to be sort of foam gumboots, reaching to half-way up the shin, with a good inch of light, pneumatic, resilient padding all round.

But remembering trout-fishing days, it's always the inexperienced, self-fancying twerp who turns up on the bank with masses of expensive, brand-new equipment which he proceeds to show he doesn't know how to use. The true veteran brings only his old, well-worn gear and his hard-acquired skill. Still, I couldn't very well produce experience of the Antarctic which I didn't possess. All I could do was pick Eric's brains and get what he reckoned I'd need to see me through. Eric was in London and therefore accessible on a telephone. Ronald Lockley was in New Zealand, and therefore much less so.

It was an exceptionally forward January in the Isle of Man. The daffodils were well up already, several in bud; snowdrops, aconites, winter heliotrope and viburnum fragrans all in bloom. It would be summer in Antarctica, of course (no snowdrops), and Eric had assured me that it wasn't all that cold all the time. The really troublesome feature, he added, was the catabic wind. Catabic? *Shorter Oxford English Dictionary*: 'Affording an easy descent'. The antarctic continent rises high above the surrounding antarctic ocean, and consequently there is a steady spill of cold air (convection currents) off it, which pours over the ocean in the form of 'catabic winds'. (Strictly speaking, surely, it's the continent, rather than the wind, which is 'catabic'? Never mind.)

'But what the hell d'you want to go to *Antarctica* for?' asked the boys in the local. 'Why not the Caribbean, or India, or somewhere nice? Not much fun in Antarctica, I wouldn't have thought.' 'Birds,' I replied. 'It's one of the greatest areas for bird-watching and wildlife in the world. Also, very few people have been there.' 'Ay – and very few'd *want* to, I reckon. Have another pint? No draught beer where you're going, you know.'

Ever since we became friends, in 1973, Ronald Lockley, a fine ornithologist if ever there was one, had often spoken to me of the phenomenon known as 'The Great Antarctic Food Circle'. 'Those lonely, cold southern oceans are teeming with food – krill – for sea-birds, and they're not disturbed there. That's the place to see the oceanic sea-birds in their own wild haunts. It's a bird paradise, totally unspoiled.'

'You've been, Ronald?'

'Alas, no, I never have, and now I'm seventy-seven. But damn it, I'd still go if only I could! There's an organization called Lindblad Travel . . .'

There was. There is. And Penguin Books, consulted, thought that Ronald and I might very well write a book about our travels. What about a photographer, though? We'd need one, to help produce a book. I co-opted my friend Peter Hirst-Smith, a cheerful, easy-going harlequin with a camera. So oceanic sea-birds, here we come, armed with the eyes, the companionship and the expert knowledge of Ronald Lockley and Eric Hosking! Eric was sailing as an ornithological lecturer on the *Lindblad Explorer*'s five-week voyage, starting from the little port of Ushuaia in Tierra del Fuego, down through the Drake Passage to the Bellingshausen, Amundsen and Ross Seas, and so up to Stewart Island at the southern tip of New Zealand. 'If you go out watching birds with Ronald,' I had said again and again to my friends, 'you see five times as much as ever you'd see on your own.'

What books should I take? I wondered. This has always been a tricky question for me, as for so many people setting out on a long trip. In the end I took the *Odyssey*, Shakespeare's sonnets, selected poems of John Clare, Yeats's poems, the second volume of Boswell and Ronald's *chef d'œuvre*, 'Ocean Wanderers'. Weight was the problem, of course, and I didn't dare risk more. Armed with these and with Jane E. Brody, I packed and set forth.

Twenty-four hours later I found myself in Terminal 3 at Heathrow, fidgeting about impatiently (and quite unjustifiably) while poor Peter (actually *wearing* his fluorescent orange waterproof jacket) paid £160 excess baggage fare on his photographic equipment. The writer has only his pen by way of technical equipment, but life's awkward for a photographer. A cup of tea, some farewells and then we were through passport control, through baggage check, through the departure lounge and aboard Argentine Airlines' jumbo jet, for an eighteen-hour flight to Buenos Aires. There we should meet Ronald, who was coming the other way, from New Zealand; and Eric and Dorothy Hosking, who had preceded us via New York.

I have no enthusiasm for long flights on aeroplanes. What a pity you can't become unconscious – just hang yourself up, like a coat in a cupboard, until you're wanted again! Although I'd specifically checked with the Argentine embassy in London that we didn't need typhoid injections, Peter had had one all the same; so he couldn't even have a drink, poor chap.

Three and a quarter hours out from Madrid; and quarter past one in the morning. Looking out of my window on the port side, I recognized several northern hemisphere constellations. To an English eye they all appeared strangely angled. Cassiopeia was standing up like a capital sigma, Perseus lay horizontal and Andromeda was plunging vertically down the sky to disappear over the horizon. Pegasus was not visible, but the Gemini were tilted diagonally sideways. Soon, however, cloud blotted out this extraordinary show and I went to sleep.

On long flights I never alter my watch until actual arrival, since not doing so enables you to tell more easily how long you've been going, and also the duration of the flight. At eight o'clock the next morning, English time, it was still, of course, pitch dark outside, South American time being five. We were awakened for toy breakfast. ('Plastic breakfast,' said Peter.) As nights on jet planes go, I'd had a good one, and felt quite fresh (though it didn't last). There ensued the usual scrummage for the loos. ('Scusate!' 'Pardonnez-moi!' 'Frightfully sorry, madame, je suis busting...')

As light stole into the sky we found ourselves cruising high above a level floor of flocculent, dappled cloud, stretching as far as the eye could see. Three and a half hours to go.

We came in to Cordoba for a forty-five-minute stop. This, my first sight of South America, was rather an anti-climax: a green, flat chessboard of fields, nothing more. Dawn, however, was a deep-orange, glowing gold, like a furnace door opened in the east. Then the sun came up. Two minutes from the first glimpse and there it was – a complete, circular disc. How fast does the earth revolve, how far away is the sun and what is its diameter? How long it takes to rise is obviously both inevitable and calculable, yet the speed always surprises me.

The flat airfield and the start of a hot summer day. A warm, humid smell of vegetation and a dark-blue, jagged, Van Gogh-like line of hills along the skyline. Argentinian air steward (brusquely): ' Go inside!' Well, he's got virtually no English; probably doing his best. Even finding an imperative was no doubt quite a strain. I obeyed, and returned to Dr Johnson. I wished he could have been on this trip. ('Penguins, sir, cannot but excite our risibility by their resemblance to human beings. A crowd of aldermen on an ice-floe is a sight so unexpected as to compel us to smile at its very absurdity.')

Between Cordoba and Buenos Aires we flew over a huge plain – illimitable, it seemed – totally flat; cultivated, green-and-brown fields, little squares between perfectly straight roads intersecting at right angles. The dwellings lay among patches of trees – planted for coolness and shade, I suppose. Across this symmetrical, artificial landscape, the courses of rivers wriggled like worms – the only irregular features to be seen.

Landing at Ezeiza airport, Peter and I made the acquaintance of another Lindblad traveller – one who promised (and later turned out) to be rather good value. This was a lady aged about sixty-eight; a Miss Hartwell. Miss Hartwell was about five foot two, dumpy, fresh-faced, retaining excellent teeth, energetic, very direct and English in manner. She was travelling by herself and clearly disposed to be friendly. On the way in to Buenos Aires by taxi, it transpired that she was a very experienced traveller indeed – a dedicated globe-trotter. I quite literally could not find *anywhere* that she had not been to, except St Helena and Ascension. (Yes, she *had* been to Easter Island; and to Madagascar; and to Ulan Bator; and the Aleutians.) She had been travelling the world for thirty years. She must be very well off, thought I; though her clothes were not particularly expensive and her manner was ordinary English upper middle-class. 'Where do you live when you're not on these amazing travels?' I asked. 'In a hotel in Belgravia.' She was travelling first-class on the *Lindblad Explorer*, and after this Antarctic trip was going straight on to Peking and Lhasa.

Having reached the hotel, Peter and I began looking out for Ronald. However, by half-past two he still hadn't shown up, and as by this time we were both rather drowsy after our night on the plane, we went to our rooms and slept for two or three hours. At six o'clock that evening, as we were crossing the hotel lobby to go out for a walk, round the end of a cucumber frame, whom should we meet but Mr McGregor! He was standing at the reception desk in a flowery Hawaiian shirt. There is always something very heart-warming about the mere sight of Ronald, seventy-seven and looking about sixty, imperturbable and exactly the same whatever part of the world he happens to be in. He was annoyed at having forgotten to bring his copy of *South American Birds*, and at once began telling me about some owls he'd heard calling by night in Santiago. They must have been hunting pigeons, he thought. He wished he could have identified them.

After we'd all had a drink, Ronald showed me a copy of his latest book, *Endangered New Zealand Species*, beautifully illustrated by Noel Cusa. I can't remember ever to have seen better bird and animal illustrations. They were in watercolour – reproduced, of course. The reproduction of colour was excellent and Ronald was pleased that in nearly every case Cusa had depicted the birds actually doing something authentic.

And here, if only space allowed, I would go on to set out in detail what a horrible mess has been made of the endemic and indigenous species of New Zealand: indeed, of the whole ecology of that beautiful country. Unthinking slaughter; and many harmful predators irresponsibly and thoughtlessly introduced from England, etc. For some species it's already too late; for others it may be. What a piece of work is Man!

Later that evening we went for a stroll in the Plaza Britania. I was delighted to see a statue of George Canning, calling the New World into existence to redress the balance of the Old for all he was worth. So they still remember! I've always thought that one of the most enjoyable things about travelling is to come upon bits of history on the spot – actually to see local evidence of events and things which hitherto you've only read about. To have a Danish friend show you where Nelson's ships lay during the bombardment of Copenhagen – to stand on the Heights of Abraham, where Wolfe fell at Quebec – to see the place where Cook landed on Papeete – this is the pay-off for having had to learn some history. (A statue of Louis Kossuth stands in Cleveland, Ohio; there being a strong Hungarian enclave there.)

Canning's statue in Buenos Aires (opposite)

LA
NACION
ARCENTINA
A
EORCE CANNINC

*The seeds of the
turtle tree*

Otherwise, the Plaza was rather dull – mostly plane trees and
acacias. But we saw a very small pigeon – the smallest I've ever
seen – which Ronald identified as the mourning dove. We also
came upon a scattering of fallen seed-cases, bipartite, convex,
black, very hard and about as big as folder book-matches. These
were the seeds of the so-called turtle tree. The cases do indeed
very much resemble the shells of those little turtles they sell
(damn them) in aquarium shops.

Dinner in the hotel restaurant. The waitress, a charming girl
called Katerina, took a fancy to Peter and knocked 50,000 pesos
off the bill – the equivalent of about £11. (I mean, she made the
bill out for less than it should have been.) She told us that her
mother had been a Lithuanian gipsy and her father a Hungarian,
and she certainly looked up to all of that. It's strange how many
quite ordinary people – not travellers, but people with their
livings to earn – wander about the world, far from their birth-
places, and are to be found working at relatively low-paid jobs
in unexpected places. We went to bed in excellent spirits, full of
Katerina's buckshee red wine.

Next morning we met another English spinster who had
booked on the Lindblad trip. This lady, Miss Betty Robinson,
was altogether different from Miss Hartwell, though about the

Identifying flowers in Buenos Aires (Ronald on right)

same age, or even a trifle older. She was slight and thin, bespectacled, frail in appearance, with a marked Cockney accent. She was very chatty and full of enthusiasm. I formed the opinion that for all her frail physique she would be dauntless when we got into the cold and ice. (And so it transpired. *Vide infra, passim.*)

A second walk in the Plaza Britania that morning proved more rewarding than that of the previous night. Peter photographed first the seeds and leaves of the turtle tree; then a very large hibiscus tree in pink bloom; and finally what Ronald and I believed to be a species of wattle, covered with yellow flowers. Familiar as I am with the hibiscus in various guises, I'd never before seen one in the form of a tree fully as big as a large horse chestnut.

The estuary of the River Plate at Buenos Aires is immense. You can't see across it.

During the afternoon the three of us went on a 'bus tour of the city, but this was disappointing, except for the flowering trees – jacaranda, hibiscus, oleander and others unknown to me. The town has little beauty: and the weather had turned thundery and humid, which didn't help. True, there is no smoke or air pollution in Buenos Aires, but the rubbish piled everywhere fully makes up for that: also the garish advertisement hoardings clum-

sily topping the tall buildings. The decaying harbour area of La
Bocca was not in the least picturesque; merely ugly and depress-
ing. There may be beautiful and exciting things to be found in
the city, but it seemed to me that one would have to live and
work here, and get to know some local people, before one could
expect to strike reality.

That evening the Lindblad tour officially began with an inau-
gural dinner at the hotel for all those who were coming on the
trip; and here I at last met Eric Hosking and Dorothy. My
ornithological ignorance soon became plain enough, and indeed
I made no pretence to anything else, being more than content to
listen and learn. We spoke of Transjordanian Petra, which, as it
happens, I reached during the war. Eric told me that once, while
flying in a helicopter near Petra, he had spotted a lammageier's
nest, with young, in a high, rocky cave below. 'What,' I asked,
'*is* a lammageier, please?' Eric looked puzzled. 'But you must
surely have *heard* of a lammageier?' 'I fear not.' 'Oh. Oh, I see.
Well, it's a large bird of prey found in the Orient; as big as or
bigger than a condor. It lives largely on bone marrow, and on
that account often takes its prey to a height and drops it, to
break the bones.' 'Its prey?' 'Some people say it flies close to
goats on precipitous places and pushes them off. I have an idea
that the lammageier may be at the back of the old tale of the
ancient Greek Aeschylus, on whose head a tortoise fell.' I feel
that Sir John Mandeville could hardly have bettered the lammageier, and I hope very much to see one, one of these days.

At dinner another character made himself known, in the person of Mr Bob Leslie; Strine, ocker, crew-cut, shirt open upon a
hirsute chest, sporting a walking-stick; a retired farmer from
Adelaide; a most refreshing and invigorating character, with
some fine throwaway comments on the passing scene. His presence did me good, for I confess that at breakfast next morning,
which took place at 4 a.m. in the hotel coffee-shop, I began to
feel more like Evelyn Waugh every minute – a reaction shared
by Peter and Ronald. It was an excellent breakfast – scrambled
egg, bacon and (oh dear!) garlic sausage; all piping hot. But
Americans at 4 a.m. are apt to be a bit much for the average
Englishman. As we sat down with our trays, a sharp-featured
Texan lady of about forty looked up and said crisply – nay,
interrogatively – 'Good mawnin' Ah'm Carolahn Best.' (Like
that.)

'How do you do?' I managed. Ronald said nothing, but Peter, bless him, managed to be a little more forthcoming. Soon she asked, 'Are y'all three English?' Now, however could she have known that?

The real veterans – Eric and Miss Hartwell – seemed in excellent form; self-possessed and cheerful as we embussed for the airport in the humid, rainy dawn. The American gladhand in charge, one Mr Thomas, was efficient and helpful. By half-past six we were in flight for the Rio Grande.

A nice American couple from Seattle, Washington, sat next to me on the flight; a Mr and Mrs Frederiksen. Mr F. proved to be a civil engineer from the California coast – evidently highly successful, yet quiet and modest. All of us, at this stage of the trip, were like new boys at school or recruits to the army. We were cautiously getting to know our companions and forming provisional ideas of those we reckoned we were going to be able to get on with.

There were eighty passengers on the trip altogether. Just about the right number, it seemed to me.

We flew over the Straits of Magellan. From above, the sea looked blue and smooth; the land dark, empty, uncultivated and forbidding in appearance as we landed at Rio Grande.

'Please remain seated until the aircraft has come to a complete stop.' They say it here too. This solecism now appears to be virtually world-wide. Pray, what is an incomplete stop?

Now followed a four-hour 'bus ride from Rio Grande, down through Tierra del Fuego to Ushuaia; a matter of 150 miles or thereabouts, over narrow roads which were certainly not without their local idiosyncrasies – some of them quite startling.

At first our way lay across a featureless, bare, sodden plain of brown grass: no flowers and no birds except a few gulls – *Larus dominicus*, the southern black-backed; a rather nasty, predatory bit of work. It was mournful scenery. Tidal inlets, isolated pools of water: rather like Connemara, but even more wild and lonely. Here and there stood a few groves of the beech-like *Notofagus*. All had many dead branches and were hung with grey lichen, like trees under a curse. On our left lay the expanse of the south Atlantic, very smooth beyond a low-tide shore of rocks and pools.

As the monotonous journey continued, we did begin to see a few birds: the caracara, a scavenging hawk, recognizable by its noticeably red tail; the giant petrel – a bird with which we were later to become more than familiar; and some kind of pipit which not even Ronald could identify precisely. After a time the desolate plain gave way to a rolling terrain of open grass, interspersed with woodland, resembling a kind of bleak, untended park.

We stopped for lunch at a surprisingly good restaurant on the shores of Lake Fagnano (fifty miles long). It seemed to be in the middle of nowhere and I couldn't help wondering what – or who – on earth kept it going. There was no one there except us, but as we were over eighty strong the management were happy enough. Then followed a slow climb into mountainous country – precipitous, wooded hills rising above steep gorges and lakes below. Here the commonest birds were black-and-white martins, continually flickering back and forth across the road as they hunted for flies. Patches of snow were lying and now, for the first time, we felt a cold wind, harsh and penetrating. No doubt about it: we were a long way south, and getting souther every minute.

At last, in the early evening, we came down out of the mountains to Ushuaia, and there – how marvellous! – lay the *Lindblad Explorer* herself, awaiting us in the harbour. She – all neat red and white paint – was much the most attractive thing in sight. Beyond lay the little township, a straggle of huts and sheds scattered along a barren shore at the mountains' feet: a church, a dock, a couple of small hotels. One couldn't help thinking it a dreary, bleak place. And this was late summer. I wouldn't care to winter in Ushuaia.

We boarded the *Explorer* in good spirits; tired out (we'd been up since 3.30 a.m.) but glad to be out of the hard-seated, swaying 'buses, and eager to get aboard and have a look at the ship and our quarters. And now there occurred, for Ronald, Peter and me, a really splendid turn-up for the book. An empty first-class cabin was going spare, and Big Dave McTavish, the so-called 'expedition leader' – a husky young Australian, with whom I was later to become on close and friendly terms – offered it to me. It was a double, so Ronald and I moved in together, leaving Peter in solitary state in the 'B' deck cabin which the three of us had been expecting to share. This happy piece of generosity on Big Dave's part made all the difference to our trip for, as we quickly realized, we'd been a bit over-optimistic in supposing

Ushuaia harbour

that the three of us would be comfortable in one 'B' deck cabin. Like Hamlet, we should certainly have lain worse than the mutines in the bilboes.

At dinner that evening I met Keith Shackleton, a noted ornithological painter and one of the numerous staff of specialist lecturers on board. (An attractive feature of *Lindblad Explorer* trips is the programme of lectures by experts on ornithology, oceanography, the history of antarctic discovery and so on. Often these are eminent people; for example, among those who have lectured on the *Explorer* are Sir Peter Scott, Bengt Danielsson, Roger Tory Peterson, Thor Heyerdahl and Captain Cousteau.) It was Keith who, later, pointed out to me a condor (*Vultur gryphus*) – the first I had ever seen – sweeping across the sky above the bay. This is the Andean condor (as opposed to the Californian, which is now one of the world's rarest birds). The normal wingspan is about nine to ten feet, and to me, that evening, it looked colossal. (I had not yet become familiar with the great albatrosses.) However, within the hour Ronald spotted a yearling wandering albatross (*Diomedea exulans*) sitting on the sea no more than a hundred yards from the ship. To a mere northern hemisphere bird-watcher like me, used to nothing bigger than fulmars in the way of tube-noses, the 'built-up' beak looked enormous.

Ronald and I went to bed full of a sense of excitement and anticipation. To-morrow we would be at sea, south of Cape Horn and speeding towards the Great Antarctic Food Circle.

2

FROM USHUAIA TO KING GEORGE ISLAND

About seven o'clock next morning I woke in great comfort in our windfall cabin to find Ronald already stirring, brilliant sunshine outside and gliding through it giant petrels, their smoky-dark wings widespread as they soared and swooped around the ship. Through the porthole I could also see various passengers already walking on deck, among them Mr Bob Leslie, unmistakable, with his crew-cut and southern Strine tan.

Dressing hastily, we went and knocked up Peter to go out and photograph Cape Horn as it receded to the north. The Horn is not at all majestic, as the Cape of Good Hope is – just a lump of grey rock, really. Still, it *is* Cape Horn, and we felt it ought to be recorded as an important landmark in our journey.

Betty Robinson was on deck, beaming with warmth and happiness. The discovery that Keith Shackleton was on board had filled her cup to the brim. She had a real, pre-war-style schoolgirl crush on him. She told us that she had already seen a shoal of dolphins – 'black, dear, but white underneath, you know.' The giant petrels had been taking biscuits from her hand, she went on. (Throughout the trip Betty continually fed bread or biscuits to every living thing she could, including the monstrous elephant seals.) She certainly had a lot of nerve to try it on with giant petrels, I reckoned. These birds are called 'Stinkers', partly because, feeding as they do on sea-carrion, they smell nasty at close quarters, but mainly because if disturbed on the nest, etc., they spit a foul-smelling liquid at the intruder. They are big – nearly as big as a small albatross – with heavy, tube-nosed beaks. The wing-beat is rapid, but mostly they glide and soar on outstretched wings. Many were following the ship that morning.

Giant petrel or
'Stinker' (Macronectes
giganteus)

Oddly, they reminded me of elephants. Their eyes are small and
inconspicuous. The young bird is very dark, almost black, but
as they grow older they turn grey, and the plumage – becoming
sparser, perhaps – is remarkably reminiscent of the cracked,
fissured and folded skin of elephants. Also, they have a kind of
heavy, slow deliberation, even in flight. They sometimes sit on
the water, but not often. Their life-span is thought to be something like fifty years. Like all petrels, they lay only one egg a year
(incubation period forty-eight days!) but don't mate until they're
about eight or nine years old. They nest in small colonies – about
ten or twelve nests – on islands and large rocks. The very severe
weather conditions in which they often find themselves keep
their numbers steady – that and the skuas, which take weakened
or injured birds. Their northern range is about the north of
Patagonia, and up there they are dark-plumaged. The further
south, however, the lighter the plumage, and in extreme southern
latitudes, near the Pole (as we were later to see for ourselves),
completely white ones are not uncommon.

Coming aboard the night before, our friend Miss Hartwell
had stumbled and hit her head. I found her sitting in the lounge
with an appalling contusion on her right temple, as big as the palm
of your hand; horribly swollen, and criss-crossed with bloody
scratches. Protruding grotesquely, it looked like a 'bump on the
head' in a children's comic. Miss Hartwell, however, made light

of it. 'Nothing at all. Doesn't hurt a bit. Soon be gone.' Talk about No Nonsense!

At breakfast everyone was cheerful and talkative. Dorothy Hosking was very smart in a speckled grey trouser suit, collared and belted, half-length sleeves, with a russet jersey underneath. During the whole voyage she remained, I thought, consistently the best-dressed lady aboard.

Returning on deck I at once saw, some way astern, two black-browed albatrosses (*Diomedea melanophris*) and a storm petrel (*Hydrobates pelagicus*). The albatrosses were pure white about the heads, shoulders and under-parts; black backs and wings; and tremendously thick, curved beaks. The storm petrel, smallest of the tube-noses, fragile-looking and a dark, smoky colour all over, except for a white rump and yellow foot-webs, was clipping the waves, reminding me of a swallow skimming over the Kennet at home, sipping and splashing in flight. They seldom, it seems, actually approach close to a ship except in bad weather. The sight of these beautiful birds, plainly at home and living their accustomed lives far from any land, excited me, and for the first time I began to realize what sort of an ornithological voyage we could expect.

We were now on a south-south-easterly course from Cape Horn (we had left Ushuaia by the Beagle Channel – good old Darwin!) towards King George Island in the South Shetlands, and the Bransfield Strait beyond.

The next birds to turn up were magellanic penguins, comfortably swimming through the vast, bitter ocean. Penguins are often to be found swimming in groups far from land. They cannot fly, of course, so they are immersed for weeks at a time, and this gives some idea of their resistance to cold. Snow will not melt on a penguin. They are totally insulated, retaining all their natural body-heat.

On this, our first morning at sea, we were all given a general briefing by Big Dave. One thing he stressed – and this, I think, deserves to be recorded – was that no litter whatsoever, not even a match or a fag-end, was to be dropped either on the ice or the land. In the Antarctic – catabatic wind or no – litter never decomposes or disappears. The *Lindblad*'s rubbish – apart from that edible by birds, etc. – was always sunk, in deep water.

Everyone going ashore, Dave continued, was always to take his personal disc from the disc-rack in the waist, and return it when he came aboard again. In this way, if the ship needed to put off in a hurry, e.g. if the weather changed suddenly, they could tell whether everyone was back aboard. 'And if ever you're out on the ice and you hear a succession of short blasts on the siren, come back at once, toute suite, and the tooter the sweeter. No fond farewells to the penguins.'

As is generally known, penguins are not afraid of man (though later, in New Zealand country, we were to find the yellow-eyed penguin an exception). However, man's greater height, we were told, tends to make them nervous. They probably register man as a bigger penguin. So the proper way to get on terms with a penguin is to get down to it – kneeling or squatting.

During the afternoon of this first day out, the ship rolled a fair amount (we were going through the notorious Drake Strait) and some of the weaker brethren began to feel the strain. Cocoa was kept on tap in the Penguin Room, as the lecture-room aft is called. Cocoa, we learned, is quite an effective palliative against sea-sickness. I never found out why, but it certainly served us well.

All three of us – we'd already become known as the Gang of Three – volunteered for the Whale Watch. Throughout the ship there was an enormous enthusiasm for whales and the captain, Hasse Nilsson, was in effect a fanatic. He was reputed to be better than any other captain at sea at getting close to whales without disturbing or frightening them. Throughout the hours of daylight a continuous watch – two volunteers at a time – was kept for whales. All watchers were issued with a copy of a *Brief Guide to the Identification of Cetaceans in the Southern Hemisphere*. This, while almost incredibly badly written, did give enough information to enable amateurs to identify any whale, dolphin or porpoise they were likely to clap eyes on. This included details of six species of rorqual, two species of right whale, three species of sperm whale, two species of porpoise, fourteen species of dolphin (including two species of killer whale) and eight species of beaked whale.

There are, in fact, about seventy-six species of whale, porpoise and dolphin in the world altogether. (If you want to get particular about sub-species, there are considerably more.) These are divisible into two principal orders: *Odontoceti*, or toothed whales, which includes porpoises, dolphins, sperm whales and beaked whales; and *Mysticeti*, which includes humpbacks, right whales and rorquals.

The *Mysticeti* are baleen whales. They live on krill (*Euphorzia superba*); masses of little crustaceans, about two inches long, with which these cold southern oceans teem. The 'baleen' consists of fringed plates or 'combs', depending from the roof of the whale's mouth, the function of which is to filter krill and other food from the sea. This flexible baleen is the so-called 'whalebone' formerly used for corsets and so on.

Rorquals have folded or pleated throats – throat-grooves, which are really short and numerous baleens. They are fast in the water, and until the introduction of the harpoon gun, in about 1865, were on this account killed less frequently than right whales.

It is the right whale which has been hunted to extinction, except for a few sub-species such as *Eubalaena australis*, the southern right whale. It was named the 'Right Whale' by the old-time whalers, because from their point of view it *was* the right whale to kill. It basked on the surface and was a slow swimmer. It was unafraid of the approach of man. Indeed, it virtually offered itself to the harpoon. It couldn't sink, even when dying. It contained thirty tons of oil and had very long baleen plates. To kill two was enough to make a voyage profitable. This is why there aren't any left.

On our second morning out from Cape Horn we woke to find the sea calmer and the rolling of the ship abated. In fact we were lucky in this part of our voyage, for the Drake Strait, between Cape Horn and the South Shetlands, is rough and wild as a rule. We crossed it under exceptionally calm conditions.

Coming on deck, we found that the giant petrels had left us. All we had now by way of bird companions were the little, dark-plumaged storm petrels darting just above the calm sea.

Mr Frederiksen, the Californian engineer, wishing to say something genial, asked me whether I was Sir (*sic*) Richard the Lionheart. 'No,' says I, 'more like Richard the Third, I fancy.' This meant nothing to Mr F., but nevertheless his good nature remained unabated. As might be expected, everyone aboard was thoroughly amiable nearly all the time. For one thing, they were on holiday, but over and above that a lot of us, I think, felt rather devils to be up for such a trip at all. Our English spinsters, Miss Hartwell and Miss Robinson, did England the greatest credit; always imperturbable, self-possessed, cheerfully on top of the world. They seemed like the English maiden ladies abroad of a hundred years ago, of whom one reads. I could just imagine Miss Hartwell, for example, in the Indian Mutiny; while Betty was the very image of the golden-hearted Cockney – full of fun at seventy-odd, always giving or taking a joke, and up for P.T. at seven-fifteen every morning. (So was Ronald. So was I, reader, come to that.) I hope this English archetype never disappears, for it commands universal liking and respect.

There was still very little wind and we were beginning to wonder when we would start encountering it. The prevailing winds of the southern Atlantic are the 'roaring forties' – westerlies. But we had now passed through their latitudes virtually unscathed. (From Cape Horn down to the South Shetlands you pass from about latitude 55° to 61°.) However, there is also a belt of easterlies along the coast of the continent itself, and these were still to come.

During that second day we saw no less than eight species of sea-bird: wandering albatross, black-browed albatross, giant petrel, blue petrel, magellanic diving petrel, storm petrel, magellanic penguin and sooty shearwater. The first time you see one of the big albatrosses – royal, wandering or black-browed – you wonder whether you're dreaming. The wingspan is well-nigh unbelievable. I don't know whence Coleridge got the idea, but I now realized for the first time the true power of the image. You'd have to be an awful swine deliberately to kill one of these. They have enormous grace, majesty and dignity, and with this goes a kind of detached beneficence. They inspire awe and delight. (The awe comes from realizing what they are capable of doing and withstanding, as a normal part of their lives.) An angel could do no better than look like an albatross, really. We were happy to watch them for half an hour at a time – or longer – as they followed astern, back and forth, their great wings spread motionless to the wind.

During the next day the ship sailed uneventfully on, and early in the afternoon passed the convergence of the antarctic and sub-antarctic waters and entered the antarctic ocean. The temperature of the water fell by 3° to 2° Celsius. The weather became still colder and now there was a cutting wind from ahead on the starboard side. At about half-past five, as Eric Hosking was in the middle of a lecture on bird photography, he was interrupted over the intercom. Whales had been sighted close to the ship and the captain was closing on them and altering course to follow.

Everyone dashed out on deck. To describe the excitement of the next two hours is altogether beyond me. This was one of the biggest moments of the whole voyage; and, I think I would add (just speaking personally) of my life. I'm fairly sure that most of the other people aboard would agree. We were among a herd of perhaps forty finback whales (*Balaenoptera edeni*) – the second largest whale in the world. They were each about sixty feet long, in colour a uniform dark-grey, with a greyish-white patch on the belly. They were feeding on krill and small fish, hunting through the upper waves and surfacing to blow about every half-minute. At times we would find ourselves within twenty-five or thirty yards of three or four whales, all surfacing at once. Again, we would see ten or twelve surrounding the ship. They paid not the least attention to us, although they must, of course, have been able to hear our engines and been aware of our position in relation to themselves. As a whale rose, one saw first the white

The finback whales. 'You could have thrown a cricket ball onto the backs of some of these whales'

spray as it blew, then the huge curve, beginning at the crown of the head (the nose itself seldom appeared) and the convex, dimpled double orifice of the blow-hole, a foot in diameter. As the blow-hole disappeared, the length of the back curved up and over, forty feet from behind the head to the erect triangular dorsal fin far down the body, near the tail. The triangular tail – or 'fluke' – itself seldom broke surface, unless the whale happened to turn in its course as it went under.

From the deck you could have thrown a cricket ball onto the backs of some of these whales, so close they rose all round the ship. The ponderous, unhurried smoothness with which they appeared, curved over and vanished below the sea gave a tremendous impression of power and strength. I could never have imagined anything to compare with this reality: it was like a vision. The ship's officers told us that they themselves had never before seen a herd – or 'pod' – of whales anything like so numerous: nor at such close quarters.

Also attracted by the krill were hourglass dolphins (*Lagenorhyncus cruciger*); longitudinally black-and-white striped, each about six feet long. These were diving in and out of the waves in the wake of the whales. Various birds were feeding, too – little Wilson's petrels, wandering and black-browed albatrosses and black-and-white mottled pintados, or Cape pigeons. The petrels flew just above the waves, their legs extended to patter along the surface as though walking. (Petrels are so called after St Peter: *v.* Matthew XIV, 29.)

After about two hours the ship resumed her course for King George Island. All of us shared a common feeling of delight from this remarkable experience. The atmosphere on board was rather like that of a wedding, a victory or a declaration of peace. Everyone was exhilarated and on voluble good terms with anyone else at all, and for a long time that evening no one had a thought or a word of conversation for anything but the whales.

Opposite, top: *Big Dave piloting a zodiac*
Bottom: *the* Lindblad Explorer *at Ushuaia*

This splendid pod of fin whales may have been near its present southern limit of ocean-roaming. To-day it is said to be less polar in its range than the blue and minke whales, preferring deep water from sub-polar up to sub-tropical seas.

It has up to 114 grooves or pleats extending from the chin to the navel, enabling the throat to expand to enclose up to a tonne

Right: *iceberg (with penguins) in the Bellingshausen Sea.* Below: *the* Lindblad Explorer *at King George Island.* Bottom: '*Many of their sloping faces were stippled*'

of water and food in one closure. Compressing the pleats with powerful muscles, it expels the fluid through the baleen plates, of which up to 350 hang from the 'gums' of the upper jaw. These plates, which vary in length – longest in the middle of each side of the upper jaw (up to three metres or ten feet in the blue whale) and progressively shorter towards the hinge and tip of the mouth – are tough, flexible and fringed with ragged hairs, ideal as a sieve for retaining solids. They are not in the least like teeth, but are horny and grow continuously, like human nails, as the extremities wear away with use.

In A Whale for the Killing, Farley Mowat describes how the fin whale homes upon a shoal of fish by echo-location, transmitting a pulse of low frequency which enables it to identify the prey it prefers. (In his account it was herring.) 'Closing with its target, it begins circling the school at torpedo velocity of twenty knots. Now it sways sideways to present its white belly to the school, a flashing ring of reflected light.' Once the fish are concentrated in a tight bunch, scared by the white flash of the whale's underside and flippers, the finner expands the accordion of its huge, pleated throat and sucks in the ball of fish. The tongue rises to expel the water and collect the food, which is rapidly digested. When feeding a big calf during the southern summer, the blue whale mother produces enough milk to increase the average weight of her child by some 90 kg (200 lbs) daily. It might be supposed that this would be a severe strain on a lactating female, but in fact she too puts on weight daily, engulfing the many tons of krill and small fish necessary to lay up fat for the long voyage to wintering grounds in warm tropical or sub-tropical latitudes.

I myself have seen, in the Alaskan fiords, the same pattern of rounding up planktonic and small fish prey. These fiords are the summer home of the humpback whale. This slower-moving, chubbier whale emits a ring of air bubbles from its twin blow-holes under water. The reflected light from these bubbles and the white underside of the whale's very long flippers scare the milling ball of krill to the surface. The whale then rises vertically beneath. The enormous gape sucks in the food; and sometimes hovering gulls, snatching at the popping shrimps and alevins, are drawn down into the vortex and swallowed whole. This phenomenon gave rise to the erroneous report in earlier accounts that this whale included sea-birds in its diet.

Top: *Blue-eyed shag on nest*
Bottom left: *Gentoo penguin*
Bottom right: *Elephant seals moulting*

The finback whales. Probably a cow and calf (Balaenoptera edeni)

In the confused movements of the pod of whales which we were following I could detect no pattern of such rounding up. Probably there was no need for them to do more than plough through the diffused banks of Euphorzia krill and engulf what they needed at will.

We had been lucky in meeting such a concentration of these large whales. We were to see a few other whales, but not the majestic finner again. It was a vivid lesson in how easy it has been for these friendly leviathans to be slaughtered by man. They allowed the Lindblad Explorer to cruise right in among them, within easy harpoon shot.

The breeding biology of the great whales is well-known from studies carried out by cetologists at flensing stations where they have been cut open and their fat and bones converted to marketable oil and other commodities. Of the ten baleen species, one is now extinct, while the largest – the blue *Balaenoptera musculus* – survives only precariously. Perhaps 1,200 of these now exist, chiefly off Newfoundland. By estimation there used to be around 100,000 in antarctic waters in the summer, but now they are rarely recorded there.

The gestation period is just under one year. At birth, in the tropics, the calf (rarely twins) is estimated to weigh over 5 tonnes *and measures 7 m (23 ft) long. It has little or no fat when born (hence the need for a warm tropical nursery). However, it quickly acquires a thick coat of blubber and, a month later, accompanies its mother on the spring migration to polar seas.*

The calf is weaned in the autumn, at seven months, at which age it is about 16 m (52.5 ft) long. Up to seven years the baleen plates grow with a distinct pattern of annual ridges. The age of any immature whale killed during that period can thus be accurately assessed. At four years (22.5 m or 74 ft long) only four per cent are sexually mature. Growth continues up to the tenth year, by which time the blue whale is physically mature and up to 30.5 m (100 ft).

The blue whale mates during its sojourn in tropical waters, the female being ready to receive the male not many days after giving birth to her calf. She is unlikely to conceive during the long lactation; lactating cows taken in polar seas are usually found to be 'empty' (the whaler's term).

Judging by the slow rate of recovery of the blue whale population since it was afforded total protection in 1966 by the International Whaling Commission, the mature female probably does not calve more often than once in two, or even three years. Although she may mate each winter in the tropics, she needs at least one year 'out of production' to recover from nursing so huge a calf.

Similar statistics apply to the next largest, the fin whale. (This is approximately one-fifth smaller than the blue.) From an estimated half a million before whaling began two hundred years ago, fin whale numbers have dwindled to around 150,000 world population. Although now given total protection by the I.W.C., the fin whale is still hunted. Several maritime nations are not members of the I.W.C., although world opinion and the serious decline in whale breeding stocks are factors encouraging them to join. Some ships, flying 'flags of convenience', take any whale they happen to find. Legally, such pirate whaling ships may not operate within the newly-established 200-mile limit claimed by maritime nations around their coasts. However, there are as yet no such limits in Antarctica, and later on our voyage we were to see at least one Japanese whale-killing ship among the icefields.

3

KING GEORGE ISLAND

At quarter to six next morning I was woken by Ronald digging me in the side. 'Icebergs, Richard! Icebergs!' I jumped out of bed and looked out of the porthole. Sure enough, about a mile away to port lay a very large, flat-topped, bluish-tinted iceberg, looking like a gigantic, white table; the first I had ever seen. It must, I suppose, have been a good half-mile long, but of course I couldn't make any estimate of its diameter. Other, smaller bergs were lying all around, from sizeable islands down to lumps of floating ice.

I clambered into my thermal underwear, fisherman's sweater, ditto socks, balaclava and all the gear which I had so painstakingly bought in the Isle of Man. 'So this is it – the Antarctic!' Everything felt fine except the plastic mittens-over-gloves, which were clumsy and made writing difficult. (We were both taking notes everywhere, all the time.) I took off the mittens, and out on deck we sallied. You could feel the incredible, bitter cold kept at bay outside the clothing; a reassuring and pleasant sensation. The first person we met was Mr Bob Leslie, in jersey, cotton trousers and *bare feet*! 'She'll be right, mate. I've worn bare feet all me life. Goin' to try fire-walkin' in Bali one of these days.' Betty Robinson was walking thirty-eight times round the ship. 'Thirty-eight times is a mile, dear. Keith told me, so it must be right, mustn't it?'

On our starboard side lay King George Island (discovered in 1819, so George III); a rocky mass, ridged and snowy, though here and there the dark, bare land showed between the expanses of snow. It was about three miles away, and we could plainly see the whole curve of Admiralty Bay. There were no traces of habitation. All about lay the icebergs, one with two sharp twin

cones several hundred feet high. The sea was calm and the sun very bright. The scene was idyllic – and colder than anything we had ever known.

A few giant petrels were flapping about close to the sea's surface, for since there was no wind they could neither glide nor soar. Plenty of Wilson's petrels were skimming round too, but no albatrosses. Further out, perhaps four hundred yards off, we glimpsed some creatures which I thought at first must be more black-and-white hourglass dolphins, leaping with their curving motion in and out of the water. However, they turned out to be penguins. Soon we realized that there were other groups of penguins on all sides of us; some 'porpoising' through the water, others swimming on the surface. On the surface they rather resembled very large guillemots, raising their long, pointed beaks vertically into the air whenever they took a look around them. They did not approach the ship. These were either Adélie or chin-strap penguins – two smaller species common in this latitude.

There are seventeen species of penguin in the world altogether, and these are divided into six genera. Only three of these genera – the *Pygoscelis*, *Eudyptes* and *Aptenodytes* – include antarctic species. In all, we were to see twelve species during our voyage.

Admiralty Bay
(pages 38–9)

King George Island

King George Island is a good sixty miles long and perhaps twenty miles across. The highest point is only 2,228 feet, but the numerous 'nunatacs' (an esquimau word, imported to the Antarctic, meaning 'bluffs', 'headlands' or 'high points') give it a rugged, mountainous look. Admiralty Bay cuts about ten miles into the heart of the island.

As we were preparing to go ashore we saw the grey shape of a seal lying stretched at ease on a small berg drifting past. It gazed at us impassively, lifting its head vaguely but not troubling to move. It was a Weddell seal, Leptonychotes weddellii. This species ranges circumpolarly on both fast and pack ice. Huge concentrations gather to pup in the spring after a winter spent largely under *the ice. Here they maintain breathing holes by chewing the ice with their small, hard teeth as fast as it builds up and freezes the surface. In the darkness under the ice-sheet, this amazing seal finds its fish food by echo-location. Scientist-divers, overwintering in Antarctica, have swum with Weddell seals and recorded their underwater moans and 'songs', trying to interpret their significance. Dominant males, signalling their presence vocally, apparently patrol submarine territories, jealously on guard during the mating season.*

The Weddell female is slightly larger than the bull. She suckles her child for only six weeks, at the end of which time it is enormously plump from living on milk which is more than fifty per cent fat – 'liquid butter', as it has been described. The thick covering of blubber beneath a dense fur coat enables these antarctic seals to endure midwinter temperatures down to minus 50° Celsius. When it gets colder than that, they haul out to sleep beside their breathing holes. It is, of course, warmer to stay in the sea beneath the ice, but not so restful; seals can sleep for short periods under water, but must half-wake when they rise to renew the oxygen in their bodies by gasping air at the surface. Under water the heart-beat is reduced to about ten per minute by a special sphincter muscle which controls the flow of blood. At the surface, while breathing air, the normal pulse rate is up to 160 per minute.

The record seal dive seems to be held by a wild Weddell seal studied by Dr Gerald L. Kooyman, who fixed depth gauges on some individuals and recorded descents to 250 fathoms

The four crab-eater seals (Lobodon carcinophagus)

(1,500 ft = 457 m). *One Weddell seal remained submerged for twenty-eight minutes between descending and ascending; if the whole period was spent in diving and coming up again, the rate of travel would be roughly one foot per second, but allowing for a feeding period of half that time of twenty-eight minutes, the speed of travel must have been double. Exactly how fast a seal descends we don't know, but at a depth of 457 m the pressure of water is 700 lbs (317.15 kg) per square inch, which is forty-six times that of the atmospheric pressure we live in at the surface!*

Next we saw, about fifty yards away, four crab-eater seals floating along on a small berg about as big as half a lawn tennis court. This creature, Lobodon carcinophagus, *is altogether different from the dark-coated Weddell seal; more elegant – if that is the right adjective – sleek, smooth and supple, in a cream-grey fur coat with coffee-brown streakings, and about 8 ft (2.4 m) long.*

Its head is more dog-like, with a longer nose. It is more of a pack-ice dweller, with teeth conspicuously lobed for draining water from mouthfuls of krill, its principal food. It weighs about 4.5 cwt (225 kg). The total antarctic population is high – at least ten million.

The crab-eater seals watched us closely and warily. They certainly had something to watch. You never in your life saw such a sartorial bunch as the eighty passengers of the *Lindblad Explorer* ready to go ashore! Any cartoonist would have thrown in his pencil and owned himself beat. People were padded out like Mr Michelin; sun-goggled, face-creamed, moon-booted, encased in brilliant scarlet, orange, yellow, bright blue. There were two little Hawaiian-Japanese boys, brown-faced and very American, who all through the trip seemed to spend most of their time scrapping with each other. This morning they looked rather like miniature returning astronauts, in baseball hats, goggles, plastic red windcheaters and knee-boots. But they were as nothing to some of the adults. Like the Ancient Mariner's, we were a ghastly crew.

We were taken ashore in the *Explorer*'s zodiacs – the first time most of us had had any experience of these extraordinarily rugged and practical craft in action. A zodiac is about fourteen feet long and shaped, in ground-plan, exactly like a flat-iron, with a pointed bow. Upon the vertical, wooden stern is clamped an outboard motor. The sides, which at the forward end curve inward to meet at the bow, are made of inflated, rubberized canvas, very tight, smooth and convex, rising a foot or so above the flat interior well. Up to about sixteen passengers sit upon these sides, facing inward. The well comes in handy for dumping clothing and equipment (especially when you're going scuba-diving). A zodiac has virtually no draught at all – perhaps an inch or two – and in capable hands can go safely among rocks and into caves where a plastic or wooden boat would be in danger. This is because the rubberized sides are pneumatic and resilient. Internally, they are divided into seven separate compartments; so if one of these were to puncture (which I never saw happen) you'd still have another six to keep the zodiac afloat.

The ship had some six or seven zodiacs in all. They were stored one on top of another in two piles on the upper after-deck, and launched and hoisted in again by a special winch.

They made all the difference to the voyage. By their means we could, except in rough seas, go ashore almost anywhere, steer through and over brash ice with safety, enter deep caves where the swell was tossing against the rock walls; cruise round the rough coasts of islands; and approach birds, seals and any other sea creatures we wished. In a calm sea they could be taken two or three miles from the ship with perfect safety, and often were. Whenever we spent time ashore in any particular place, a shuttle service of zodiacs would be maintained to and from the ship, from the time of landing until the time of departure. Their only fault that I could ever observe was a tendency on the part of the outboard motors to 'play up' – stall or refuse to start at awkward moments. I treasure a memory of Big Dave at Campbell Island (of which more anon) apostrophizing a temperamental outboard which kept stalling. His language startled even me. The ladies pretended not to hear (but people got used to Dave's language anyway). All the *Explorer*'s staff – Dave, Keith Shackleton, Jim Snyder, Tom Richie *et al.* – could handle a zodiac with skill and precision. It was, in fact, required of them as one of the qualifications for the job.

Watching us as we came ashore was our first penguin on dry land – an Adélie (*Pygoscelis adeliae*). This is the species which most closely resembles everyone's mental picture of a penguin: white front, black wings, back and head; no splotches or white markings anywhere. (Squeak, of Pip, Squeak and Wilfred, was an Adélie.) They are about two feet high and on each hind flipper are three long black claws. This fellow was unafraid; merely, perhaps, a shade more wary than a pigeon in Trafalgar Square. He let me come to within about five yards and then walked away, hobbling in a comical but dignified manner. It is virtually impossible not to anthropomorphize penguins, pop-eyed in their dress suits, walking upright and appearing so extremely self-possessed. They walk clumsily, but once they take to the water swim with grace and ease. 'Pinguis', of course, means 'fat', and although they don't look fat in pictures, when you actually see them they are in fact very rotund behind and before. It's all insulation. The water they swim in is usually about 30°F. (The ocean freezes only at about 28°, or even less than that.) Later that morning we came upon some gentoo penguins. These are about the same size as Adélies, with red bills, orange legs and a white patch across the top of the head.

Adélie penguin
(Pygoscelis adeliae)

The ground underfoot was stony for the most part, but there were large areas of a short, velvety moss unknown to me, and large clusters of *Usnea*, a lichen with a golden tinge and jet-black apothecia, which is very common as far north as the Falkland

Beach on King George Island

Admiralty Bay
The 'windrow' of the
blue whale
(Balaenoptera
musculus)

Isles. There was also a monocellular, crimson alga growing thickly *under* patches of old snow. The effect was to tinge wide areas of the snow red in a very striking way, rather as though it were bloody, or – to be more exact as to the shade – stained with something like crushed strawberries.

A little way along the shore was lying the entire skeleton of a blue whale, killed long ago by the whalers. I paced it out. It was sixty-three feet long, so the whale would have been about seventy-five feet, I suppose. The bones were immensely thick. These great skeletons, when they lie on shore, are called 'windrows'; an expressive and rather emotive name; but on looking it up in the *Shorter O.E.D.*, I found that the only definition of the word was 'a row of mown grass or hay'. So it's evidently been 'applied', metaphorically, to the whale skeletons.

I lay in the warm sun while Peter took photographs and Ronald looked for petrels' nests; but found none. All of a sudden we were startled by a growing roar, like that of approaching aircraft coming in fast. We all looked up in astonishment, but

the noise was not caused by aircraft. A mile off, across the bay, a glacier along the shore had 'calved'. In other words, the seaward end of the glacier had broken off, dropping as a great mass of snow and ice into the sea, to become an iceberg. The air all along the cliff was dense with white, misty snow-powder, and we could see it continuing to fall, *slowly* as it seemed, into the heaving, turbulent water, long after the roaring sound had ceased.

> And through the drifts
> The snowy clifts
> Did cast a dismal sheen –

But in fact the ice has an amazing, luminescent tone – either blue or green – as though containing fire or light. The reason is that the hollows and recesses fill with sea-water, until the crevice walls become packed with it. One is, therefore, looking at a substance which is not merely ice and which has a refractive index different from that of ice – one which emphasizes the blue end of the spectrum. Certainly the visual effect – a caerulean glow deep within the ice – is one not easily forgotten. If there were a jewel with such colouring and luminescence –

I saw a huge berg, far off, the centre of one flat end of which consisted of a great, round-headed opening, perhaps fifty feet high. It could not have looked more symmetrical if it had been man-made – a perfect Norman arch. 'Would there be an ice-floor inside?' I asked Keith Shackleton. 'No; water.' 'Like Fingal's Cave?' 'Perhaps. But anyone who took a boat in there would be asking for trouble. These bergs are likely to split and break up at almost any time.'

Above the shore of Admiralty Bay, side by side, stand four big, wooden crosses surmounting cairns of piled rocks; memorials to four young men, all in their early twenties, members of the Falkland Islands Dependency Station who died here, two in 1948 and two in 1959. 'In an accident' one memorial reads. I didn't inquire about details. Poor lads!

Later on, during the afternoon, we saw two humpback whales at a distance of about four hundred yards from the ship. They were just idling along, doing nothing much and blowing now and then. As often as they rose, they displayed their great flukes as they went under. This, of course, is the picture which most people have in their minds when they imagine a whale rising to blow. The huge size of the fluke is very impressive.

Here, too, we saw skuas for the first time. The skua is neither a beautiful nor an attractive bird. It is both a predator and a scavenger, eating both live prey and carrion. It preys upon anything wounded, weak or helpless, particularly deserted or neglected chicks of other birds. In colour it is brown, with white wing markings; as big as a raven, with a nasty, pitiless appearance and a powerful, hooked beak. There are two species in this region, the antarctic and the brown; but Eric Hosking told me that they can be distinguished only in the hand. As the voyage continued and I became familiar with the ways of skuas, the very sight of them used to give me the shivers. Whenever you came upon anything exciting or beautiful, there always seemed to be a few skuas hanging about, waiting to pounce (in which activity, as will be narrated, they were not infrequently successful). Skuas are very good at their job.

During the afternoon the ship crossed the bay to the neighbourhood of the Polish Research Station, where we went ashore again. Our zodiac passed some rocks on which was a colony of blue-eyed shags. These are about the same size as the European shag, but of more attractive appearance. They have black heads, backs and wings, white breasts and under-parts, and the epony-mous – and very conspicuous – sky-blue circular eyelids round their dark pupils. Between and slightly above the eyes are two caruncles or fleshy lumps, dull yellow in colour and rather like lichen in appearance. The birds' general gait and behaviour was much like that of European shags.

The shore was littered – thickly scattered, indeed – with windrows of the bones of whales; a dismal sight. Foraging

Skua (Catharacta maccormicki) *and chick*

among these were two or three giant petrels, very tame. Ronald and I walked up to within a few feet of one, which stayed put while he photographed it. Further on, we encountered a herd of elephant seals; enormous beasts, the females fully as large in body as an English cow, the males about one and a half times that size. They were lying on the stones, huge and gross, pressed together in groups of six or eight, simply basking in the sun, which was now very warm – as warm as a fine May day in England. They were moulting, and their chocolate-coloured fur and skins were coming away piecemeal, leaving ugly, bare patches. They seemed torpid and showed no fear of us, except when we approached very close – within two or three feet – when they adopted threatening postures, opening their mouths wide (a cavernous red, with short fangs not very visible in the gums) and, as it were, offering to attack. As part of this display they roared (or mooed – somewhere between the two), but they were so bulky and ungainly that at six feet away we felt safe enough. They were lying in masses of their own excrement and smelt worse than any zoo. All the bulls had the great bulbous, ridged noses, like small trunks, from which the creature gets its name.

Clambering on, we came to a big rookery of gentoo (*Pygoscelis papua*), Adélie and chinstrap (*Pygoscelis antarctica*)

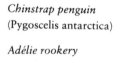

Chinstrap penguin
(Pygoscelis antarctica)

Adélie rookery

*Chinstrap feeding
chicks*

penguins. All these Pygoscelian penguins are much the same size, and similar in behaviour though not in markings. The rookery covered the crest and slopes of a low, rocky hill above the shore. It comprised tens of thousands of birds, among which we wandered as we pleased, while they simply went on with what they were doing. This consisted principally of feeding their chicks, which at this late stage of the breeding season were well grown, fully as big as their parents and considerably fatter – or they looked it, anyway, being covered with fluffy, grey down which would later give place to their adult, black-and-white, closer-fitting plumage.

The feeding process struck me as an extraordinarily inefficient business, after millions of years of evolution. A female parent which had been off to sea and filled its crop with krill would return and call raucously to its chick. About three chicks would come running and she would then try to discourage two of them by swearing and pecking, while simultaneously trying to encourage her own. As they were all equally indefatigable, this often proved a failure, and after a bit the wretched mother,

Gentoo feeding chick (opposite)

Gentoos and whale vertebra

harassed beyond endurance by the wrong chicks, would end by taking to her heels, lurching from side to side as she ran. The chicks pursued her, but a wrong one usually seemed to prove the fastest, so that the whole pecking and swearing business would begin again. It all seemed rather pathetic. Eventually, however, the mothers did manage, of course, to feed their own young. As they opened their beaks wide, the chicks would push their beaks in and gobble up the regurgitated krill.

I estimated the area of the rookery at about three or four acres, but could form no estimate of the innumerable penguins. Dotted over the area was a number of dead birds, both chicks and adults – the work of the skuas. The skuas, rather bigger than ravens, passed and re-passed in continual flight at very low level above the rookery, going fairly fast in twos and threes and plainly looking for something to attack and kill. The penguins simply ignored them – what else could they do? – but all the same, I can't help feeling that a penguin knows quite as much as it wants to about skuas and lives in continual consciousness of them. They reminded me of the German bombers in France in 1940: nothing much to be done except hope it isn't you.

Among the myriads of penguins wandered occasional sheath-bills; white, ugly birds, about as big as chickens and not dissimilar in gait and behaviour, with strong, thick, triangular beaks. They can fly if they are put to it, but in practice spend most of their time wandering about in penguin colonies, eating carrion and excrement. They, too, will seize helpless or dying chicks when they can. The penguins never attack them, however, and don't even seem to object to them.

The smell of the place was strong and after a time made one feel slightly nauseated. However, there were compensations. We watched, at a distance of about five feet (the penguins would let you do anything except actually touch them) a pair of adult chinstraps making up to one another; as it were, renewing their courtship display. Standing close together and face to face, they lifted their heads vertically and made raucous calls, at the same time caressing each other with heads and beaks. It was long past mating time, yet their mutual attraction was plain. They were simply affirming the existing bond between them, as Ronald assured me. It is their way.

The rocks were thickly covered with beautiful patches of golden and orange lichen, rather like our English *Xanthoria aureatum*, but even more vivid in hue. The sea inshore was very clear and clean, with always the same caerulean, green hue shining out of the ice.

At the end of the afternoon, while re-boarding a zodiac to return to the ship, Eric Hosking slipped and fell into the shallow water. He was none the worse apart from a wetting, but unfortunately he wrote off one of his beautiful cameras beyond repair.

It had to be explained to a few among us, who thought these ragged-looking individual penguins were sick and dying, that they were on the contrary in normal health. Before beginning the moult all penguins lay on extra fat, to tide over a fast of up to one month. In a study of the gentoo at Macquarie Island, Pauline Reilly and Annie Kerle give a time-table for the moult of this penguin. During the week of what they call 'premoult', the body weight increases by up to forty per cent through heavy feeding. The bird now stays ashore. During the next three days the tail feathers fall out first, while the flippers swell and are liable to bleed if accidentally struck against an object (as when the bird is handled for weighing). The body-feather moult begins on the tenth day after coming ashore. It is very rapid, the bird standing upright but rather hunched up and in a comatose condition (like some we saw that day on King George Island). Provided the moulting bird is not disturbed by human activity, the new feathers emerge and gradually push out the old, which at first 'stand up', largely concealing the new growth. If the penguin is compelled, by human disturbance or a summer gale, to move about, the loose feathers fall or are blown off, giving it that ragged appearance which led our friend Betty Robinson and some other tender-hearted visitors to pity it.

But it is all a perfectly natural process; the moult is completed usually within three weeks and the penguin is once more waterproof and eager to get at the swarms of krill in the sea a few yards away. On average it loses, during this fast, 3.3 per cent of body weight per day. 'Mean weight loss of twenty-four adult gentoos during 18 days of moult was around 50 per cent; from 8 kg it fell to 4 kg in that time.' So good is the feeding, however, that on returning to the sea, 'one bird, 4.2 kg when released, had gained 1 kg when recaptured 3 days later'.

4

DECEPTION ISLAND: AND A REMARKABLE ICEBERG

At eight o'clock the next morning the ship passed through the narrows known as Neptune's Bellows and entered Port Foster in the centre of Deception Island.

Deception is an almost circular volcanic crater, lying at the southern end of the South Shetlands chain. It is shaped like a South Seas atoll – a horseshoe, with a lagoon in the middle. The only entry from the outlying ocean to the central lagoon is inconspicuous compared with the size of the island as a whole. The average breadth of the horseshoe is about two and a half miles, and the island's circumference is about twenty-one miles. The early sealers, looking for a natural harbour, at first thought there was none, until at last they happened on the 600-yard-wide strait known as Neptune's Bellows. Within lies the empty, central lagoon, known as Port Foster. This is a mile or so across and very calm and sheltered, though the cutting wind (catabatic) sometimes whips it up.

Deception is an active volcano, with hot springs and fumaroles. There was an eruption in 1924, when the lagoon literally boiled and the sulphur and the heat together stripped the paint off every ship in the place. There were further eruptions in 1967, 1969 and 1970. We kept our fingers crossed. The island is an example of a *caldera* – a volcanic complex, the upper part of which has collapsed along ring faults to form a large interior basin of subsidence.

Since the previous, sunny day the weather had changed for the worse. As we came out on deck – Peter with his camera protected in a plastic bag – we encountered drifting fog, visibility down to about 600 yards, bitter cold and the sharpest wind yet. The *Explorer* anchored and we saw the usual landscape of bleak slopes, patched with expanses of thin snow. The highest point of the island is only 1,806 feet. It is called Mount Pond. This reminded me of the Red Queen in *Alice*: 'I could show you hills, in comparison with which you'd call that a valley.'

Zodiacs were lowered and we went ashore. Both the beach and the steep slopes leading off it consisted of black, volcanic, finely granulated lava, which crunched and slid underfoot – awkward going. Climbing proved difficult. You went two steps up and one back. It was best to hold on, as best you could, to some vertical rock wall close at hand, and then heave. There was a tearing wind. I had a thick balaclava *and* my hood up on top of that. (Keith Shackleton, I observed, was bare-headed: a fine piece of lifemanship, but I can't imagine how he stood it.)

The first creatures we encountered were seventeen bull fur seals lazing on the beach. Among them were walking the usual snowy-white sheathbills, pecking away at what there was to peck away at. (You're never short of a meal if you're a sheathbill. Harry Wharton's favourite gibe, 'Yah! Go and eat coke!' would

Deception Island. Fur seals (Arctocephalus australis) *and sheathbills* (Chionis alba)

merely be taken literally: the sheathbill would accept with alac-
rity.) The seals were tawny and black, with bulging necks – fat
and fur – a kind of lion-like ruff all round their heads. This
serves as a protection when the bulls fight, and also comes in
handy in a mating display. These seals certainly were on the
pugnacious side. One came for us most aggressively; and we
retreated, for Ronald warned us that they can inflict a nasty bite.
This seal, however, merely saw us off and then returned to his
comfortable lazing in the driving wind and sleety rain which had
now come on.

This is the widely distributed fur seal of the southern hemisphere,
Arctocephalus *species. It is not a true seal, for true seals have
trailing legs and are almost helpless on land, hunching along by
lifting with the fore flippers and throwing the body forward.
The fur seal is one of the* Otariidae *or eared seals, and is in fact
a sea-lion. It has small, external ears, and is able to turn its hind
flippers forward as feet, on which it can walk quite rapidly –
even canter – also using the long front flippers, quadruped
fashion.*

*The difference between the eared fur seal and the true seal was
nicely illustrated that morning when we encountered a crab-
eater seal lying near the fur seals. It looked up at us placidly but
did not shift its pale body, lying supine with rear flippers encased
in flesh and fur, almost a miniature whale's tail. This tail it uses
to propel it in the water by a side-to-side vertical stroke. (The
whale's tail, by contrast, is moved up and down on a horizontal
plane.) Probably the aggressiveness on land of the fur seal has to
do with its gregariousness and the bull's fight to win a harem
during the breeding season. The crab-eater, on the other hand,
seems to be a loner when it comes ashore, either on ice or beach,
always close to the sea. It appears to come out of the water only
to rest, and probably when in fear of attack by its principal
antarctic predator, the orca; and, as with this individual we saw,
when moulting – which is a period of fasting. Both seals and
penguins lay up fat for the moult, and at this time avoid sub-
mergence in water, which would chill the skin during this 'un-
dressing' period.*

By contrast the leopard seal, Hydrugra leptonyx, *is a lean-
looking, almost slender animal, with a large head and huge jaws.
Males can reach a length of ten feet (three metres): females are*

smaller. They are solitary animals, wandering far during the early adolescent years, although normally resident in Antarctica and sub-antarctic islands as far north as South Georgia. Occasionally an individual will reach the northern coasts of New Zealand. Once one turned up on the sandspit below my Auckland home. It was typically dark-grey above and lighter grey beneath, liberally spotted with the darker, leopard-like rosettes which give the animal its name. In character, too, this seal is like a leopard, stalking its animal prey – small seals, penguins and fish – with silent and sinuous motions of its long body, sneaking through the water, in and out, without splashing.

Leopard seals mate in the southern antarctic spring, during November and December, soon after the females have given birth to their single pup on the pack ice. As with all true seals, the lactation period is short, lasting only a few weeks, but the pup is very fat and can stand a long fast before it learns to fish for itself and begin the solitary life of all leopard seals.

A steep, rocky slope led us up to the nesting sites of several pintados or 'Cape pigeons' – a sailors' name. They are actually fairly small tube-noses, mottled black and white. It is true,

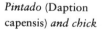

Pintado (Daption capensis) *and chick*

however, that the general impression they give is rather like that of pigeons. The fluffy, grey chicks were sitting beside their mothers on the ledges. One chick and its mother both spat their pink, krill-stained fluid at me. Many tube-nosed species make use of this deterrent, and very effective it is. 'Don't you get the stuff on you,' said Dave. 'You won't get it off in a hurry, and it's a lot more noticeable than Chanel, I can tell you. No one'll want to know you for weeks.'

We scrambled still higher, up to the rocky notch called Neptune's Window. This reminded me of Dow Crag in the Lake District; a high precipice, mist swirling over its steep faces, with a splendid outlook over the beach below and across the iceberg-strewn sea beyond.

Along the beach, as we returned, antarctic terns were flying – the first I'd ever seen. To me, terns are the most attractive of all birds, and on this voyage we were to see many southern hemisphere species new to me – including black ones. There was also a Weddell seal – just one – mottled brownish-grey on the back, with a lighter-coloured face and pale, creamy-grey underparts; a very beautiful creature. Ronald, who lives for seals, was delighted.

An exciting activity had been planned for the afternoon. It was intended to take the ship down to a place called Telefon Bay, where we were again going ashore to see the volcanic craters. Later, anyone who wanted to would be able to swim in the fumaroles, or hot volcanic springs. I was looking forward to this, since I love outdoor swimming and wanted to add a bathe in Antarctica to my 'list' (Europe, Africa, Asia and the two Americas). But it was not to be. The met. was duff. In other words, the wind, rain and fog increased and the captain decided that we would have to leave Deception for the open sea. So out through Neptune's Bellows we returned, and southward aye we fled.

During the evening the ship passed to the south of the Liège Islands, and in this sea we came upon great numbers of icebergs, some very large indeed. One in particular excited overwhelming wonder in everyone. It is difficult to convey the impression it made, for this was due in large part to the oceanic solitude, to its size and our proximity to it. It bore an uncanny resemblance to a man-made castle, all shaped in luminescent ice. This, of course, is the game that we play with clouds, or with 'pictures in the fire'. If there is an infinite number of plastic shapes, whether of vapour, of glowing coal or of ice, some will, fortuitously, resemble things we know.

> Sometime we see a cloud that's dragonish,
> A vapour sometime like a bear or lion,
> A towered citadel, a pendent rock,
> A forked mountain, or blue promontory
> With trees upon't, that nod unto the world
> And mock our eyes with air.

This iceberg was in the form of a Norman keep. The symmetry was most strange – almost disturbing. It was about eighty feet high and I would say roughly the same cubic size as the keep of the Tower of London. Its square turrets surrounded a flat, smooth, almost regular top. Below, in the centre, an arched entrance opened upon a central courtyard of ice. The deep soffits of this arch were evenly striated, a glowing turquoise, and outside, round the threshold, lapped an emerald-green sea, foaming

in shallow waves. The whole great, floating island, as it drifted, was utterly silent, and round it flew the white snow petrels, about as big as doves. (*See cover.*)

It is easy enough to say that of course all this was mere coincidence and fancy – a game of the imagination. This was a large lump of floating ice. In another few hours its shape would have changed and the likeness would have vanished. Believe me, reader, if you had stood on deck in that freezing solitude, gazed up at those glassy, blue towers and through that icy arch, and reflected that it was only by mere chance that any human eyes happened to see it at all, you would find it (as we did) very hard to dismiss lightly. It was like a castle deserted and bewitched – all turned to ice. As the ship approached and slowly passed it close to starboard, we all stood amazed. And so it floated on, into the falling darkness and the waste of empty sea, while we held on our course. Even so did *Voyager II* pass Saturn and continue on its way. I shivered, and returned to the lit saloon for some trivial reassurance – ribaldry and alcohol. Here the captain told me that in all his experience of the Antarctic he had never seen an iceberg which had made so deep an impression on him.

Ronald, who had not been up as far as the pintados' nests that morning, expressed surprise that the parent birds had not left their young when we approached them on the ledges. He told me that to stay put was not generally characteristic of tube-noses, and he could only suppose that these particular birds had built up their unusual reaction as a defence against, e.g., skuas and other predators.

That afternoon, as we were leaving Deception, we had seen in the distance a ship which the officers told us was a Japanese whale-killing vessel. There was much indignation among the passengers and several of the staff too. I could not bear to think that they might be going to find our friends of the night before and kill them with explosive harpoons. Yet that was exactly what they had come so far to do, and they were fully equipped to do it. Perhaps they did in fact kill those very whales. Will nothing stop this wickedness? I wish everyone could *see* for themselves those finback whales which we saw. As with most evil, it's a case of 'out of sight, out of mind'.

Top left: *black-browed albatross.* Top right: *light-mantled sooty albatross (photograph by Eric Hosking).* Bottom left: *snow petrel.* Bottom right: *antarctic petrel*

King penguins round the ship at Macquarie

5

PARADISE BAY AND THE BISMARCK STRAIT

We woke early to find the ship cruising down the narrow Gerlache Strait, with the Antarctic Peninsula – the continental mainland – to port; and to starboard, Brabant Island. Not long after, we entered Paradise Bay – a bay of the mainland – and anchored.

Paradise Bay, perhaps, hardly merits its rather flattering name. However, it is certainly majestic, with a mountainous, austere beauty of its own. It is about five miles long and two miles across, with three entries. On its shores there are at present two scientific stations, one Chilean and the other Argentinian. Although the bay looks very open when seen on the map, from the deck of a ship within the anchorage it appears almost landlocked. It owes its name, I think, to its sheltered windlessness, which must have afforded tremendous relief to sailors in the nineteenth century – and later, for that matter.

That morning the bay was completely calm and still, and only the floating ice broke the reflections of the surrounding, snowy mountains, which in many places rose precipitously sheer out of the water. The Argentinian station was sited on a little, low promontory, where no avalanche could reach it. The air seemed as warm as summer – so warm that I went up to the bridge to read the thermometer. The temperature was 36.5°F (2.5°C). All the same, it *was* warm – simply because there was no wind. Light, semi-transparent veils of mist hung in the still air at no more than a hundred feet, partly masking the mountain-sides.

Top: *inside Captain Scott's hut at Hut Point*
Bottom: *the kitchen in Scott's main hut at Cape Evans*

After breakfast, zodiacs were launched and we went for a trip round the bay, each boat setting off in a different direction. Looking at the steep rock walls rising all round in the solitude, one lost all sense of scale until another, distant zodiac came into

view, far off across the water, full of coloured dots – bright-clad
people, minute under the cliffs. At first glance their tiny size
came as a shock.

Any number of blue-eyed shags were out fishing. I inquired
what they might be fishing for, and was told *Notathemia*, the
antarctic ice-fish, which has no haemoglobin in its blood. Later,
our zodiac came close up under a shelf on which was a nesting
colony of these birds. The nesting season was coming to an end,
and most of the young were almost ready to leave. Although
they were as good as full-grown, they were nevertheless distin-
guishable from the adults because they had not yet developed
blue eyes or the characteristic yellow caruncles on the forehead.
Some of the parent birds, however, were still sitting on their
rather striking nests. These are circular and more or less symmetrical, about two feet in diameter and nearly a foot high, so
that the bird looks rather as if it were sitting on a hat-box – a
good, solid one. Our zodiac came to within about ten yards, but
the shags were not bothered. In appearance they are more beautiful than European cormorants, their white breasts and under-
parts contrasting strikingly with their black backs.

These blue-eyed cormorants or shags (Phalacrocorax atriceps)
*had well-grown young in substantial nests built of seaweed, mud
and guano. The adults were handsome: white-breasted, black-
backed, with ivory-white cheeks, yellowish caruncles just above
the brown bill and white wing-bars. The iris is grey-brown; the
eye-ring is blue.*

*They re-build the same nest year after year, until it becomes
a high cone, rising conveniently above local snow-level when
they return to breed in the spring. Like all cormorants, they have
no bare brood-patch. They incubate with their big feet, all four
toes connected with webs and hot with blood vessels for the
purpose, as they squat on their two or three white eggs, their
legs and feet hidden under the body feathers.*

*This cormorant flies rather laboriously, as if it were in the
process of evolving towards a state of flightlessness. (This state
has already been reached by the Galapagos cormorant, which
can do no more than flap its stumpy wings.) Cormorants do not
need wings to swim with, as some birds do underwater (e.g.
ducks, puffins and guillemots). Their wings are tightly closed in*

the submarine chase of their fish food, the broad, webbed feet being twirled behind as a kind of propeller. Yet, curiously, cormorant plumage is rather loose and not particularly waterproof for submarine work. After each long feeding dive, the cormorant will perch on a prominent place in sun or wind (or both), holding its wings extended until they have dried out. It does this even in the bitter Antarctic.

Gentoo penguins were also fishing in the bay, 'porpoising' in and out of the still water. We pursued one for a short time, but he could swim faster than a zodiac going almost flat out! However, we didn't persevere: there was no point in terrifying the poor little chap. As we cruised on, the brash ice jostled and *Brash ice in Paradise* hushled continually under the boat with a peculiar sound which *Bay* I can hear still.

Paradise Bay

Iceberg undercut by the sea

The volcanic rock ledges and shelving precipices all round us were covered in layers of snow, horizontally striated. Each striation represents a year's deposit of snow. The average depth, as near as I could estimate, was about two to four feet. As for the floating bergs, many of their sloping faces were stippled – dimpled – a mackerel-sky effect which, emitting the blue luminescence characteristic of all antarctic ice, was extraordinarily beautiful.

Another creature we saw fishing that morning was an elephant seal, but he gave us a fairly wide berth. (We had become so well accustomed to antarctic creatures showing no fear that I felt mildly surprised. Perhaps he was just busy.)

Later, after we had returned aboard, I happened to be up in the bows as the ship moved gently up to a low, flat berg, perhaps twenty-five yards long, on which was lying a crab-eater seal. He was a fine specimen, a good nine feet, his sleek coat olive-green in colour. As the bows came towering above him he showed no concern whatever, merely looking up at us for a few moments and then looking away again. We called out to him, but there wasn't a flicker of response. The bows gently bumped the berg

and the ice began going off like a succession of gun-reports. Not the slightest reaction from Master Seal. Then the captain spotted him from the bridge, the ship backed off and we left him to his basking.

That evening we crossed the strait to Port Lockroy, off Anvers Island. We came in under the 5,000-foot height of Mount William, bleak, empty, precipitous, inaccessible. It put me in mind of Mulgar-Meerez (in Walter de la Mare's 'The Three Royal Monkeys') 'whose snows have known no Mulgar footprints since the world began'. Surely no one can ever have climbed Mount William, I thought. (But apparently they have.)

The object of crossing the strait was to hold a barbecue on shore. It was not, in the view of Eric Hosking, Dorothy and myself, a very pleasant occasion: standing up to the ankles in penguin shit (pretty strong), in a sharp, cutting wind, trying to eat tough, lukewarm steak either with the fingers or with fragile, plastic knives and forks, while skuas hovered attentively nearby. (Skuas can turn nasty, especially when they're sharp-set and can smell meat.) I packed it in early and came back in a returning zodiac to play chess with the doctor.

At 11.40 p.m. (still light) I was beginning to feel worried about Ronald, and going to the gangway found the last zodiac just arrived back, with him aboard.

'Marv'llsh party on rocksh! All shinging shongs to piano accordion. You sh've been there, shtupid fellow!'

I wonder whether I shall be able to do that when and if I get to seventy-seven? What a man! He'd also been walking for two hours alone, far over the high ground, searching in vain for nests of his beloved storm petrels.

The ship was impressively well run. The lady purser – a strapping, handsome Swedish girl with dark-red hair and astonishing, ice-green eyes – seemed to have the administration well under her control. Everything was clean, everything worked and yet one was seldom conscious of anyone actually working. I was *Anvers Island* particularly intrigued by the oriental stewards, mostly Hong

Kong Chinese. They all spoke English and their waiting at table
and housework in the cabins were virtually perfect. What inter-
ested me, however, was their self-containment and detachment,
and the way in which they obviously led their own personal,
dignified lives among themselves. They never chaffed each other
in public, or laughed or larked about. Silent and unsmiling – yet
obviously without any particular resentment against the Europeans; we were just what they worked with, as a farmer works
with cattle – they had the efficiency of complete familiarity,
when necessary talking quietly to one another in their own
tongue, all liquids and labiates. One could read off their names
from the discs on the board by the sally-port – Kau Tam Tai, Li
Siu Chai, Yuen Hau Kan, Chan Kai Yuen, Chan Kam Ki. (I
wonder how our European names strike them?)

Next day, ashore on Torgerson Island, in the thick of a big
rookery of Adélie penguins, Ronald and I came upon a sad –
indeed, a heart-rending sight – a typical example of Nature's
callous indifference. Thousands of penguin chicks were being
fed all over the place – mothers running about, chicks running
after them. Near our feet one small, grey, starved-looking creature was standing hunched and motionless, with closed eyes,
paying no attention to anything around it. 'Obviously hasn't had
a meal for days,' said Ronald. His mother was lost – gone – most
likely eaten at sea by a leopard seal. (Quite a number of foraging
mother penguins are.) No other female would feed him. So he
was standing there, waiting to die, with the life of the rookery
going on all round him, and other chicks full of krill. Soon the
skuas would move in, and the sooner the better for him; only it
was taking such a long time. To tell the truth, I couldn't refrain
from tears, and I don't know that I feel particularly ashamed to
admit it, either. However, there was nothing to be done: there
are thousands the same every breeding season.

On return to the Explorer *there was a discussion about how
three different species of small penguin – Adélie, gentoo and
chinstrap – are able to co-exist successfully in the same terrain.
Ecologists maintain that different species cannot co-exist except in ecological or geographical isolation. That is, two species*

cannot occupy the same territory, eating the same food at the same times of day or night and breeding at the same seasons. Competition would drive the weaker species to extinction, or it would have to change its habits; or else, possibly, coalesce by fertile mating with the stronger, to become one single species.

All three of these species of small penguin, however, breed 'sympatrically' (same time, same place), and feed very largely on the antarctic krill. Recent studies have helped to show that there are sufficient ecological differences between them to permit close association without loss of identity or territory. They are not known to interbreed. Although the Adélie prefers to nest in dense rookeries on flatter or smoother ground than the gentoo and chinstrap penguins, which during our visit to Antarctica seemed to choose ground broken by rock outcrops, all three species may be found nesting separately or together in both these types of habitat. The real differences seem to be subtle, lying in certain diet preferences. The gentoo, largest of the three, takes more fish and larger-sized krill. (Krill normally form two layers in the sea, the adults living at deeper levels than the adolescents during the day, but rising to the surface at night.) Adélie and chinstrap penguins tend to feed on the younger krill near the surface by day, the Adélie taking smaller krill than the chinstrap. If this is correct, each species forages selectively and normally at different depths. This would explain the unusual phenomenon of the apparent co-existence of the three species. Male penguins tend to take larger krill than females – as measured in samples collected from males and females returning with food to feed their young.

In all three species, two eggs are usually laid, but often only one chick is successfully reared.

The stiff tail feathers of these penguins, a few inches long, are known as the rectrices, and have the function of supporting them on land in the normal upright position. Without this 'tail', consisting of spike-like feathers like those of a woodpecker, a penguin, supported only by its two small, thick, webbed feet (small because a large splayed foot would be at risk in these regions of intense frost), would tend to fall over backwards. The loss of their rectrices constitutes a further reason for their relative immobility during the two to four weeks of the annual moult.

6

TO THE ROSS
ICE SHELF

Now began a fairly long voyage at open sea, from the Bismarck Strait to the Ross Ice Shelf. This took eleven days. I made a comparison of the distances on the globe. It was about the same distance as that between London and Gambia, but that, of course, takes no account of ice and wind (catabatic).

Antarctica is the highest, driest, coldest, windiest continent in the world. 'Make that good,' as Maria says to Feste in *Twelfth Night*. Well, its *average* height above sea-level is higher than that of any other continent. There is less liquid water on it than on any other continent. It's a polar desert, with ninety-three per cent of the world's ice. (Yes, that's right.) Its coldest temperatures are colder than anything to be encountered in the Arctic. The coldest temperature ever recorded on earth is $-127°F$ at the Russian base not far from the South Pole. And finally, the *average* speed of the wind (catabatic) in the Antarctic is 55 m.p.h.

The continent is considerably larger than the U.S.A., and the total area of the Ross Ice Shelf (which is simply empty, featureless, snow-covered ice) is about five times greater than that of the British Isles.

Blue ice, I discovered, as opposed to green, is always ice that has broken off a glacier. Water, of course, cannot be compressed, but ice can. Compression causes ice partially to melt and brings about tighter packing of the ice crystals. It is this greater density which causes the ice to refract blue light. (I don't know that I'm so very much wiser now - are you, reader?)

One of our number, Tony Irwin (a war-time Chindit and later President of the East African Wildlife Society), was standing in the middle of the penguin rookery on Torgerson Island when he opened his ciné camera, changed the film, closed it again and started filming. After five or six feet of film, however, it stopped turning, and the 'black eye' stopped too. Putting it to his ear, Tony could hear a curious sloshy, squashy noise. He re-opened it. Yes, indeed: the skua which had swooped low and unnoticed above him had done a very thorough job. In fact, Agamemnon's stiff dishonoured shroud wasn't in it with Tony's camera: and skuas are a lot bigger than nightingales.

The Antarctic Circle lies 66°33′ south latitude. At two o'clock in the afternoon of Sunday, 8 February, we crossed it. Hot mulled claret was on tap to all; and someone cleaned up thirty-seven dollars by drawing the correct time in the sweepstake.

As we forged southward, still more splendid and beautiful birds appeared, to the great satisfaction of Ronald, for they included several species which even he had never seen before. (Neither had I, of course.) The blue petrel (*Halobaena caerulea*) is about the same size as a big pigeon (in other words, fairly small as petrels go). The head has a velvety, black crown; all the rest grey on top and white underneath, with a very noticeable, square, white-tipped tail. The feet are pale blue. In flight it swoops and glides low over the waves. The antarctic petrel (*Thalassoica antarctica*) is another heart-stopper. I quote from Mr George E. Watson, *Birds of the Antarctic and Sub-Antarctic*: 'A brown-and-white fulmarine petrel, with conspicuous, dark-edged, white wing bars' (a fine sight they are, too) 'and a dark-tipped, white tail. The head, throat, back and leading edges of the wings are dark; underparts . . . mainly white.' This petrel, as I observed it, tends to fly relatively high over the water, and its wing-beats are more rapid than those of the blue petrel.

The antarctic prion (*Pachyptila desolata*) (there are seven different species of prion) is a rather compact little bird, with zig-zag black stripes (which Mr Watson calls an 'open M') along his wings and a black tip to his tail. (Well, I'm doing my best, but I'm well aware that it's much easier for a reader to look at pictures of birds than to try to visualize them from verbal descriptions.)

Sailing on towards Peter I Island, we found ourselves among a dense and extensive ice-pack. There were a few big, tabular bergs (though none to compare with the spectacular ice-castle of a few days before) but for the most part the sea resembled a huge, flat plain of ice tussocks, studded here and there with

Large tabular iceberg knolls and low hillocks. The captain adopted various, divergent courses to pass round the thicker ice-packs and avoid being beset. Indeed, at one time, for quite some hours, we were sailing north-north-east – in other words, in the opposite direction from our destination, McMurdo Sound.

As we gained greater experience of the ice and its ways, I became aware that the formation of great, round-headed, more or less regular arches in the vertical walls of tabular icebergs is a not uncommon phenomenon. The reason is that often, when an extensive 'table' of ice breaks off a glacier at its junction with the sea, the sides include crevasses (along the weaker or cracked places) in the shape of a relatively narrow, inverted V. Such a crevasse tends to be still further opened and weakened by the breaking-off of the berg of which it forms part. The ice-table, therefore, starts its independent, drifting life as a berg with one or more open clefts along its sides. On these the ocean, powerful, patient and irresistible, gets to work, hollowing and smoothing them. The prolonged, inexorable action ultimately produces

'A berg with one or more open clefts along its sides'

these symmetrical, arched openings, which finally collapse, bringing about the fracturing and disintegration of the berg itself. (Hence Keith Shackleton's warning.)

This great ice-jungle was full of life, both animal and vegetable. It is a paradox that while the antarctic continent itself is largely sterile, the ocean round it is full of nutrients which support a profuse, teeming wildlife. The tropical oceans are almost lifeless in comparison. Day by day we observed, living contentedly in the bitter cold (-4°C), a great many seals (leopard, elephant, Weddell and crab-eater), the occasional whale, and any number of fulmars, pintados, prions, petrels and albatrosses. As the ice drifted past, we saw that much of it appeared discoloured or stained brown, rather as though floods of coffee had been spilt on the snow. This staining was caused by diatoms – plant plankton which provides food for the myriads of krill. The colder the water, the higher the plankton's capacity to dissolve gases, including oxygen. The dissolved oxygen combines with such silicates as fine grains of sand to feed organisms, e.g. diatoms. This is the basis of the food chain of the Antarctic – the most prolific ocean in the world.

About ten o'clock one evening in mid-February, as I was admiring several big icebergs (whose appearance suggested that Henry Moore might have been trying his hand at making pep-

permint fondants for giants), we came in sight of Peter I Island. This island, lonely and desolate, was discovered by Bellingshausen, the captain of a Russian expedition (and a great admirer of Captain Cook) in 1822. It was, at that date, the most southerly land to have been reached by man. (Cook himself never actually sighted the Antarctic continent, though he is on record as saying that he thought it probably existed. His great maritime achievement, of course, was to prove that there was no land mass – apart from Australia – in relatively southern latitudes. This he did by circumnavigating the world on the Antarctic Circle.)

Bellingshausen was unable to land on Peter I, owing to adverse weather and the inhospitable shore, and in fact it has been landed on only twice, in 1929 and in 1948 – on both occasions by Norwegians. Its inshore waters are by no means exhaustively or reliably charted, and are always beset with a good deal of ice. We had been half-hoping that *we* might be able to land, just for the hell of it; but what with the lateness of the hour, the ice-packs, the hazardous waters off-shore and the necessity to

conserve fuel and keep to schedule if we were to reach our destination in the McMurdo Sound, the decision was taken not to
attempt it. I dare say that it may also have occurred to the
captain that if someone were to meet with an accident, he himself
might have to explain what we had been up to and whether it
had really been a necessary part of the cruise. After all, the tour
organizers had made no promise of an attempted landing on
Peter I, and he was under no obligation. Anyway, the island
looked most bleak, dangerous and uninviting; not to mince
words, a right sod. In shape it was a more or less regular, blunt
cone (like a bun), rocky, crowned with snow and looking un-
utterably desolate in the fading light. A red sunset behind it
served only to intensify its wild solitude. Ultima Thule indeed.

Next morning, Ronald and I observed something which
moved us both; namely, arctic terns flying north. These terns
were beginning their annual migration to the Arctic – the longest
migration of any bird in the world. Until comparatively recently
it was not known that during an average life-span of something
like twenty-five years, these fragile-looking, graceful birds fly
about 25,000 miles each year. During August and September
they fly south from the Arctic, and having, in about a month,
reached the area of the Cape of Good Hope (or else Cape Horn)
proceed down towards the Antarctic. Here they drift eastward,
right round the continent, on the westerlies or roaring forties,
completing their moult, resting and feeding. Along about February, 'priketh hem Nature in hir corages' to begin the flight north.
Most go back through the South Atlantic, but some go up the
west coast of the two Americas and so return to the Arctic,
where they nest and breed. (A lot of arctic terns visit the Isle of
Man, so these terns we saw south of Peter I Island could well
have been in the Isle of Man last summer and may well be there
again next summer. And *their* journey costs nothing.)

On each of eight successive nights we put our watches back
an hour. For several days we had, therefore, a twenty-five-hour
day, for we were so far south that the lines of longitude through
which we were passing were only twenty-five miles apart. (At
either of the poles, of course, you could simply walk round and
round the world.) Eventually we 'lost' an entire day, missing out
16 February altogether. (Time is, of course, a purely human
invention. Actually, there is no such thing as a 'day'; only a band
of sunlight shining upon one side of the rotating earth.)

The whole ship began to be full of static electricity. As in Denver, Colorado, any metal surface touched with the fingers was likely to give a little, prickling shock.

The nights grew shorter and shorter, until there was none. On 11 February sunrise took place at about 11.40 p.m. Sunset had taken place very shortly before that, and it was in these conditions that there occurred the rare and beautiful phenomenon known as the 'green flash'.

The green flash is more likely to be seen from the deck of a small ship than a large one. Favourable conditions are a clear atmosphere, with sufficient moisture in it to make the sunset orange, and a ship which is rolling in a long, low swell. To the observers on the rolling deck, the sun seems to dip below the horizon, but then to re-emerge; and it is in this instant – a mere fraction of a second – that the green flash appears through the waves from a long distance away. The colour is brilliant beyond description – a kind of mixture of emerald and sapphire – but it is gone even quicker than lightning, and there is no time at all in which to contemplate it. The sight, however, is one not easily forgotten.

One morning, during these days of sailing to McMurdo Sound, we saw a small pod of minke whales on the port beam. There were about ten of them. These are rorquals – baleen whales – and small, as whales go, not exceeding thirty feet. (Finback whales are two and a half times the size of minkes.) These minkes were two or three hundred yards from the ship and we didn't really see much of them except their blow-spray and their dorsal fins. The blow sends up a swift-moving, low fountain of spray (different from the finback's), rather like that thrown up by dummy bombs from aircraft in off-shore bombing practice. The triangular, black dorsal fin is nearly always visible at the same time, after which the minke usually shows his fluke as he dives. Like the finbacks, these whales took no notice of the ship, but we didn't follow them and soon they fell astern.

Known also as the lesser rorqual (Balaenoptera acutorostrata) *the minke swims at considerable speed on migration, and reaches the edge of the pack-ice early in the (north and south) polar summers. In the throat and upper belly there are up to seventy pleats. These expand when water and krill or larval fish are sucked in by the huge mouth. When it closes, the pleats contract and the tongue rises to expel the water and retain the food against the flexible barrier of up to 358 baleen plates.*

Minke whales pair in late winter in sub-tropical latitudes. The gestation period is ten months. The male precedes the lactating cow to the polar feeding grounds, the mother following slowly on account of the new-born child, which enjoys the warmer water of its nursery for a few more weeks, perhaps accompanied by last year's weaner. Sexual maturity is reached at two years, but growth continues for perhaps another six years.

Our course, now that we were once more in a clear sea, was an unaltering westward one. It was rather startling to realize that in this latitude you could, if you wanted, sail west for ever – round and round the continent. As we came on from the Bellingshausen into the Amundsen Sea, our longitude somewhere about 108°W, the petrels and pintados left us. The eleventh of February was the first day of the voyage on which we saw no birds whatever following the ship.

One thing that is bound to strike an English traveller through these waters is the silence of the tube-noses as they follow astern. There is none of the continual screaming and squabbling of your northern hemisphere gulls when following ships in the prospect

Krill (Euphorzia antarctica)

of food. Petrels, of course, when not breeding, are less gregarious than gulls, and don't feel the same need to keep in touch with one another. (Gulls are rather like Kipling's bandar-log. Petrels have more dignity.) Consequently there was very little quarrelling over the edible rubbish thrown out of the ship. Ronald told me, however, that in their breeding-grounds, where there is territorial competition and also competition for mates, petrels make a lot more noise.

As we grew better acquainted with the great ocean through which we were sailing, we realized that we were travelling, in effect, through soup. This is not really much of an exaggeration. The antarctic sea is rich in food to an extent which cannot easily be grasped by anyone who has not been there. To imagine it is rather like trying to imagine astronomical distances. To exemplify: – It is estimated that at any one time there are something like three and a half million tons of krill in the antarctic ocean. Ships have not infrequently observed, from their radar, that they were passing over miles of virtually jam-packed krill, which sometimes blocks engine-room filters solid. On Zavodowski Island in the South Sandwich group, the estimated chinstrap penguin population is fourteen million – just on that one island. What they eat is krill and nothing else. The blue whale (this makes me sad, and I'm not the only one), *Balaenoptera musculus*, the largest animal which has ever inhabited the world (length up to a hundred feet), is now extinct in the southern hemisphere. Men have destroyed it. Each blue whale used to eat about 100,000 tons of krill a year, but that didn't lead to any marked diminution of krill.

The food-chain of the antarctic ocean is a simple one. Antarctic fish are bottom-feeders for the most part, though the squid come up at night. The krill themselves eat the diatoms and the carnivores eat the krill. Phytoplankton becomes zooplankton, which feeds everything else – whales, seals, tube-noses, squid, fish and penguins.

The krill, when you actually get to see it, is rather an attractive little lad, much like a shrimp to look at. Along each of its sides are six blue, luminescent 'port-holes', which emit a glow in the same way as do the tails of glow-worms and fireflies; and it has another on each side of its head. The function of these is to enable the krill to maintain communication in their dense shoals and silent world.

We were all hoping against hope to see a Ross seal (*Ommato-phoca rossi*), but it was not to be. Ross seals (named after the English antarctic explorer Sir James Ross, who brought two back to England in 1843) are rarely seen, for they live for the most part in the midst of inaccessible areas of dense pack-ice. (Sensible fellows.) Spending much of their time underwater, they bite breathing-holes in the ice with their teeth. They are one of the five 'true' seals (*Phocidae*), the others being elephant, Weddell, leopard and crab-eater. These 'true' seals are in fact the only seals inhabiting the Antarctic. The fur seals or sea-lions (*Otariidae*) occur further north, off the continent, e.g. at Campbell Island. (See Chapter 9 below.) There are probably no more than about 10,000 Ross seals in the whole of Antarctica, and they are very local in distribution. They are the smallest of the maritime Phocid seals, being only six to eight feet long. Their life-cycle remains mysterious in certain respects; for example, breeding concentrations have never been seen. Some naturalists believe that the numbers of the Ross, which has every protection under international agreement, may nevertheless be dwindling, for reasons unknown.

One afternoon, about quarter past four, as I was playing chess with the doctor, a light-mantled sooty albatross was reported over the intercom. Joining others on deck, Ronald and I had a good sight of it, both with the naked eye and with binoculars. It is a spectacular bird, the chocolate-coloured head contrasting with a white back and grey wings. These colours are very soft – a sort of pastel. It is not one of the large albatrosses, the normal wingspan being something like seven feet.

From now on we grew accustomed to light-mantled sooty albatrosses appearing among those which followed the ship. They became my favourite albatross, on account of their striking plumage, beautiful flight behaviour and a certain 'softness' of appearance. Pure sentimentality, of course; they're as tough as any other albatross. But the black-browed albatross *looks* tough – almost brutal – on account precisely of his scowling, black brows, whereas the light-mantled sooty looks positively gentle; as though he were made of Royal Copenhagen china.

Of all the albatrosses, the light-mantled sooty has the highest *aspect ratio*; that is to say, the ratio of the chord to the span; of the breadth of the wing to the wing's length from shoulder to tip. In plain words, its wings are very long and relatively narrow. Its *wing loading*, however, is rather low. (This means that its wings, being narrow, haven't a large area in relation to its weight.) Setting aside all these scientific terms, the long, narrow wings give it the most elegant, graceful appearance imaginable.

One evening – it must have been somewhere about the middle of February – standing on deck and gazing out to sea, I felt that we had at last come to the end of the world. The weather was extraordinarily bright and clear – brilliant sunshine, with hardly a cloud in the sky. The temperature was $-1°C$, but as there was very little wind the air felt pleasant – like a fine, frosty day in an English January. It was getting on for six o'clock in the evening and we were sailing due west, with the sun still high: 18° angle with the horizon, to be precise. There seemed to be almost no birds about; only a very occasional antarctic petrel or little Wilson's petrel, far off. There was no small brash ice. We were

'The end of the world' in clear water. Everything was still. All about us, as far as to the

The iceberg at dinner   horizon and beyond, lay huge, tabular icebergs. I learned later
(the radar had shown) that there were in fact seventy-five within
a radius of twelve miles of the ship. There was no estimating
their area, but the flanks which we saw varied in length from
several hundred yards to several miles. To judge by their accu-
mulated horizontal layers, most of these were over a hundred
years old; pieces broken off the continental ice shelf. One, for
example, was an estimated 120 feet high from sea-level to its flat
summit: this means that its depth below sea-level must have been
something over 800 feet. ('Think about that,' as the American

gladhand, Mr Thomas, would have said.) On another, the sea – that calm, scarcely-swelling sea – was breaking in heavy waves along the length of a curving, recessed bay. This turbulence could only be due to a great underwater platform of ice, which was churning up the sea as the berg drifted.

Icebergs are like the stars. They exist remote, in their own places, lifeless, empty, conforming to things like gravity and heat and cold, over time-spans in relation to which a human life is hardly any time at all. They come into existence, change and cease to be, but have no purpose or meaning. However, when you've lived among them for a bit, it's actually human beings who have no meaning. Icebergs are in accord with the universe. We aren't.

Having seen on the radar two exceptionally large bergs several miles away, the captain made for them. It so happened that Ronald and I were sitting down to dinner as we came abreast of the nearer one, and the ship sailed along its 'coast' at a distance of less than 150 yards. It towered above us, the precipice of the nearer face faceted in tilted planes of ice, rough-hewn as though with some gigantic chisel and polished smooth and glassy by the wind (catabatic). At the foot, however, the sea had partly eaten the ice away, so that the cliff beetled o'er its blue-glowing, cavernous base into the sea. It did indeed seem incongruous to be comfortably seated at a dinner-table, with a cloth on it, eating soup and steak as we floated within a stone's throw of this huge solitude of floating ice; but nevertheless I couldn't feel that we were acting insensitively. I'm sure Captain Cook wouldn't have disapproved. Anyway, as far as we were concerned, the iceberg certainly dominated dinner, to put it mildly. There was a certain feeling of unreality. 'I dreamt I was having dinner with Ronald Lockley, and there was this enormous iceberg . . .'

The average age of the passengers aboard, excluding three children, was 60.78. Exactly my own age! Nationalities: American 41, English 13, Swiss 7, Australian 6, German 4, Canadian 3, Indian 1, Japanese 1, French 1.

7

THE ROSS ICE SHELF,
ROSS ISLAND
AND
CAPTAIN SCOTT'S
HUTS

The evening before we expected to make our landfall (or icefall) along the Ross Ice Shelf, we crossed the international date-line and put clocks forward twenty-three hours. At about tea-time that afternoon the ship had begun to roll and pitch quite heavily in a cross-swell from port, which sent nearly everyone, including Ronald and me, to cabins. I found I was all right as long as I remained lying down, but felt uncomfortable if I tried to sit up, let alone stand up. So about six o'clock I went to bed and stayed there.

Four in the morning and the sun well up; and I, having slept long enough, got up too, went on deck and up onto the bridge. The swell had gone and the sea was calm. The temperature was −6°C, but there was no wind. The bows, the capstan and anchor chains were covered in ice and sailors, standing below the bridge, were cleaning it off the outside of the bridge 'windscreen' with long-handled rubber scrapers. Very far off, to starboard, lay Mount Terror, one of the two peaks of Ross Island; easily mistakable, at this distance, for a patch of cloud. The young officer on watch, who was the only man on the bridge, pointed it out to me. There was no one else about at all. It felt like the Walrus and the Carpenter:

> The sun was shining on the sea,
> Shining with all his might:
> He did his very best to make
> The billows smooth and bright –
> And this was odd, because it was
> The middle of the night.

I strolled about for half an hour or so, but it was plainly too early as yet for anything to be doing, so I returned below.

By 8.45 a.m. (lat. 77°S., long. 170°E.) we had arrived close to the Ice Shelf, which stretched away in either direction as far as the eye could see. It is named after the romantic and colourful Captain Ross, the first sailor to reach it. He did so in 1841, with his two ships, the *Erebus* and the *Terror*. (It was these same ships which Franklin was forced fatally to abandon, a few years later, in the Canadian Arctic.) Ross, when he first saw the Ice Shelf, remarked, 'Well, there's no more chance of sailing through that than through the cliffs of Dover.' However, he did manage to reach true land in the form of Ross Island, which lies at the extreme western end of the Shelf. Both Mount Erebus (12,447 ft) and Mount Terror (10,579 ft) form part of this island.

I could understand what Ross meant, all right. The Shelf is about 650 miles long. Its total area is equal to that of France – all flat, empty ice, 455 miles in breadth from the ocean to the continent itself. We were about forty miles east of Ross Island. The shining, white crest of the Shelf, rising about sixty feet above the sea, formed the most level thing I've ever seen in Nature. From as near as 200 yards, the top appeared as a perfectly straight line. One could fancy it to be a rampart protecting the antarctic continent from intruders. In front, like skirmishing cavalry, rode any number of medium-sized icebergs. No, indeed – you couldn't get through that with an atomic bomb.

In places we could see blocks of ice as big as cathedrals or bigger, leaning outward from the vertical edge of the Shelf; obviously about to break off and become floating icebergs at any moment – or next year – or the year after. No telling. We kept our 200 yards distance and continued cruising gently westwards. There were no albatrosses now. We were too far south. The only birds on the wing were snow petrels, as white as the Shelf itself, fluttering along it with their pretty, dove-like flight; and, of course, the skuas, easily recognizable by their size, brown

The Ross Ice Shelf

plumage, and heavy, flapping movement. (No bird goes further south than the antarctic skua. It's even been observed not far from the Pole.) A few minke whales were blowing and sporting about the base of the Shelf, and here and there we could see little groups of Adélie penguins loitering about in flat places.

Unexpectedly, as we pursued our way westward to Ross Island, we found ourselves among a group of killer whales (*Orcinus orca*). These are not, in fact, whales at all, but large members of the dolphin family. They grow up to about thirty feet in length and have conspicuous black-and-white markings; a white belly and – very noticeable, this – an oval white patch above and behind either eye. The dorsal fin is erect and very long: in the male it may be as long as six feet. These dolphins can swim fast – sometimes over twenty-five knots, so it's said. They 'porpoise' through the water in groups, and it's a fine sight to see six or seven all 'breaching' (curvetting half out of the water) at one and the same moment. Ronald told me that the reason they are able to breach together in this way is that they can communicate underwater; and even have leaders who give orders. They travel and hunt in lethal packs, and pods numbering up to 150 have been seen.

And what *do* they kill? Well, anything they can get – seals, penguins, fish and even sick or helpless whales. They're sea-wolves, in effect, and apparently when hunting they're ruthlessly methodical. It is said that they've never attacked man, but I don't think I'd care to put it to the test.

At noon the Shelf was obscured by fog, but by about 2.30 p.m. this had cleared sufficiently to disclose the lower slopes of Mount Terror, its summit still invisible above us. Ahead lay the low, conical shape of Beaufort Island, its protruding rock outcrops visible amongst its outlying ice. Ronald was excited to think that this could very well be a nesting-ground of snow petrels. The ship approached the island and lay to off it for some hours. During this time Ronald remained continuously on deck in the bitter cold, watching the island through his binoculars. He saw a very large colony of Adélie penguins, but no other birds or animals.

Early next morning we sailed down the McMurdo Sound into Winter Quarters Bay and anchored at the ice wharf, close under Hut Point. This wharf has been constructed by means of an interesting piece of technological ingenuity. A fine spray of relatively warm water was dispersed across and above an area of the bay inshore, and this duly settled and froze on the surface. The process was repeated until an area of thick, smooth ice had formed, resulting in the creation of a wharf beside deep water. To this the *Lindblad Explorer* made fast with ropes.

Ingenious it was, I don't deny; but nevertheless, the whole wharf seemed distinctly unstable or, to employ a popular term, dodgy – not to say wonky. Not only was the whole caboodle perceptibly in rocking motion, but there were several narrow rifts, each a foot or two wide, running right across it. These we crossed by walking along ramshackle strips of wood – broken-up packing-cases or similar pieces of boxes – laid across them. As we did so the strips themselves moved gently one way and another, because of the disparate movements of the two areas of ice which they were connecting. As in a wartime minefield, the safe way was marked with flags. One of our company, however – an Indian gentleman – forgot to pay heed to the flags and promptly went into a crevasse up to his backside. I'm sorry to say we laughed. ('Oh, my Gud! It is very very – '). Later that afternoon a diamond-shaped piece of the wharf, about twenty-five feet long and situated directly under our gangway, broke off with a violent crack, plainly audible inside the ship. Fortunately, no one was near it at the time, but it goes to show that Mr Mookerjee was lucky.

A few hundred yards south-east of us, up a gentle slope, lay the American McMurdo Station. This looked very much what you'd expect – a technological slum, and surely the most un-attractive square mile in all the Antarctic. It resembled a badly kept World War II camp; an untidy settlement of metal huts, ugly sheds and larger, barn-like buildings. (Other stations con-trive to look far more pleasant. The New Zealand station, two miles away, is most attractive in appearance.) No doubt, how-ever, the American station is extremely efficient. The buildings we went into were certainly warm and draughtless, though about as homely as aircraft hangars. The summer population of this place, I learned, is between 800 and 1,200. All I can say is, they could do with a sergeant-major.

Captain Scott built two huts on Ross Island – one here, at Hut Point, as he named it, and one at Cape Evans, about thirty miles further north. The latter, being situated near waters which are ice-free in summer, formed his headquarters. It was from the smaller hut, at Hut Point, however, that on 2 November 1911, this intrepid and valiant gentleman set out, with his four com-panions, Bowers, Evans, Oates and Wilson, on the legendary expedition to the South Pole which, as everyone knows, ended in their death.

Although the hut was used later, up to the 1920's, by sub-sequent expeditions, its contents have now been restored to the appearance and general state which they presented when Cap-tain Scott and his party were using it. The number of people who visit it is, of course, relatively small, and so far no one has abused it by removing articles as souvenirs.

The morning was sunny, clear and very cold indeed – the coldest we'd struck yet. The hut stands close by the shore, on the nearer side of the point and about five hundred yards from the ship. After breakfast off we set, wearing everything we had: moon-boots, fishermen's socks, thermal underwear, windproof trousers and over-jackets, balaclavas, two pairs of gloves – the lot. As I will relate, it proved by no means too much. But first as to the hut.

Captain Scott's hut at Hut Point

A few Adélie penguins were standing about as if on watch, and a crab-eater seal was basking (basking: I *ask* you!) down by the shore. Otherwise the place was deserted. The first thing to be said about the hut is that it is really a good deal more than a hut. 'A stout, square, single-storey wooden cottage' would be a better description. It was built by Scott and his party in February 1902, during his first expedition to the Antarctic, and was also used by Shackleton during his expedition of 1907–9. It is impressively well constructed of thick deal planks, with a corrugated iron roof (four gently-inclined planes meeting in a central point), and everything bespeaks the thoroughness and efficiency of the Royal Navy. It is about twenty-eight feet square, and has along three of its sides a six-foot-wide verandah, the overhanging roof supported by stout wooden columns. This verandah, on the outward side, has vertical half-walls extending from the roof to within about four feet of the ground. The purpose of the verandah was to provide extra storage space for stuff which didn't need to be taken indoors, and the half-walls were to protect the doors, windows and walls of the hut itself from getting piled up with driven snow. In fact, the lower edge of these verandah half-walls is exactly level with the sills of the windows.

The windows are double-glazed, with a space of an inch or thereabouts between the outer and inner panes. In area they are about the same as cottage windows anywhere, and the small panes (eight to each window, I think, but I didn't count) are set in wooden frames.

Inside, the floor is simply compressed, levelled earth. In the middle of the hut is a stout kingpin, from which radiate four or five heavy, supportive rafters, sloping downward to the outer walls. There are two doors, for alternative use according to the direction of the wind. There is no interior subdivision by true walls, except for a small, separate store-room near the entrance. Subdivision is by open-work wooden partitions, the spaces between the wooden uprights and horizontals being filled either by sacking or by piled-up packing-cases containing stores.

The blubber stove   The windows being few, it is fairly dark in the hut and one

had to wait a minute or two until eyes adjusted. There was a not unpleasant smell, like that of certain kinds of shop when I was a little boy in the 'twenties; seed and fodder merchants', grocers' – butchers' even. This came from the seal and penguin carcasses hanging up in the store-room (long dried and frozen stiff, of course) and the bales of straw for the ponies. Everywhere was a kind of orderly confusion of stores – tins of Huntley & Palmer's 'Captain' biscuits (with the old, flamboyant, Persian-carpet-patterned paper), Hunter's Famed Oatmeal, Sydney's Compressed Corned Mutton, Norton's Kippered Herrings. These were all of the Edwardian decade or earlier, and their old-fashioned labels and lettering brought a lump to my throat, for I remembered them from infancy in the 'twenties. The kitchen area had a blubber stove in the floor – a shallow pit with a low surround of bricks. This had a cylindrical metal chimney; a great, heavy, iron frying-pan and some similarly old-fashioned stew-pots, dippers and so on. On the partition shelf nearby stood a packing-case stencilled 'Captain Scott's Antarctic Expedition 1910. Homelight Lamp Oil'.

The New Zealand station (ice airstrip in the distance)

An inventory of everything in the hut would be too lengthy – spades, shovels, leather straps and harness, cutlery, needles, thread and wool, scientific equipment, blankets, overshoes, compasses and so on – all the items you'd expect. Seventy years is a long time. Things come to have an old-fashioned look after seventy years. Item after item put me in mind of nothing so much as Beatrix Potter. Here were the sort of cooking utensils she used, the kind of household smells familiar to her, the groceries she bought, and all the heavy, solid gear of her day. But why should I think of her rather than of, say, Kipling or Conan Doyle? Partly, I suppose, because she was herself a housekeeper and a farmer, partly because this was a domestic interior and things like these often appear in her pictures. Yet also, I think, because of the little, dark windows, the untidiness, the earth floor and the feeling of being in a den: Mr Tod's house at the top of Bull Banks, and Peter and Benjamin, in mortal terror, peering in through the dirty glass. A half-wild place, only a step from the wilderness outside, merely approximating to safety and civilization, possessing no more than a veneer of domesticity precariously imposed on a dangerous, savage environment. ('And nothing more was ever seen of that foxy-whiskered gentleman.')

Outside, it was fantastically cold, with a strong wind blowing powdered snow painfully into our faces. (Catabatic and how.) Out of the wind, the temperature was −13°C. That's 8.5° Fahrenheit; 23.5 degrees of frost F. But in the wind, which we estimated cautiously as being at least 20 m.p.h., it was −24°C. In the Antarctic three degrees of cold are recognized – Very Cold, Bitter Cold and Extreme Cold. This afternoon, in that wind, we had 56 degrees below freezing Fahrenheit. That rates as Extreme Cold. Captain Scott and his companions, during their last days in March 1912, were faced with temperatures of well over 70 degrees of frost. I now appreciate even more vividly and clearly their fortitude and gallantry. It was in such conditions that Captain Scott, himself at death's door, contrived to write those last letters to the next of kin.

We were glad to get back on board. Nevertheless, at about ten that night the indefatigable Ronald went out again, accepting a lift from a member of the New Zealand station – a guest of the ship – to go and see Weddell seals on the ice. Unfortunately, however, he couldn't get very near them, though Peter was able to take some photographs in the unsetting midnight sun.

Opposite, top: *Mount Erebus*. Bottom: *Hooker's sea-lion at Campbell Island*

Left: *royal penguins on Macquarie (photograph by Ronald Lockley).*
Top: *Peter and friends on Macquarie.* Above left: *rockhopper penguin.*
Above right: *erect-crested penguin*

Very early the following morning the *Lindblad* sailed about thirty miles north, to Cape Evans. On waking, Ronald and I found the ship anchored off-shore. Above us, to the north-east, rose the more or less regular slopes of Mount Erebus, the highest mountain in the Antarctic; an active volcano smoking into the clear, blue sky. Its whole conical shape shone dazzling white in the sun, nowhere precipitous; and this smoothness and apparent absence of ruggedness and formidability (it's not hard to climb, I was told, apart from the cold) made it look less than its twelve and a half thousand feet. The weather was beautiful, only −8°C and not much wind. Everyone, of course, was eager to go ashore to see Captain Scott's headquarters, and by half-past eight pretty well everyone had set off in the zodiacs.

Royal albatross and chick on Campbell Island

Weddell seal (Leptonychotes weddellii)

This Cape Evans hut, sited close to the shore beside a sea unfrozen in summer – so that stores could be landed – was where Scott and the twenty-five or thirty men of his expedition lived from 17 January 1911, soon after their arrival. Scott and the

other four set out from here (proceeding first to the other hut at Hut Point) on 1 November, but the rest remained at Cape Evans until the discovery of the bodies of Scott's party in October 1912 and the final withdrawal of the expedition in January 1913. This hut, therefore, was occupied continuously for two years.

In his journal Scott wrote: 'Tuesday 17th January. We took up our abode in the hut to-day and are simply overwhelmed with its comfort . . . In a day or two the hut will become the most comfortable of houses . . . It will be a fortnight to-morrow since we arrived in McMurdo Sound, and here we are, absolutely settled down . . .

'Thursday 19th January. The hut is becoming the most comfortable dwelling-place imaginable. We have made unto ourselves a truly seductive home, within the walls of which peace, quiet and comfort reign supreme. Such a noble dwelling transcends the word "hut", and we pause to give it a more fitting title only from lack of the appropriate suggestion. What shall we call it? The word "hut" is misleading. Our residence is really a house of comfortable size, in every respect the finest that has ever been erected in the Polar regions; 50 ft long by 25 wide and 9 ft to the eaves.'

All this is perfectly accurate. The 'hut' is a rectangular, wooden barn, with lean-to outhouses and store-rooms along two sides. The beds, tables and other furniture are still there, with Ponting's dark-room (he was the photographer) and the scientists' laboratory space. (The plan of the hut and its sleeping-quarters is reproduced in *Scott's Last Expedition*.) I lay down on Scott's bed, just for a few moments, and offered up a short inward prayer of thanks for all the good his example must have done. Then I sat at the table, in the place where he sat at that last dinner party of 22 June 1911 – the one he describes in his journal. (Seal soup, roast beef and Yorkshire pud; flaming plum pudding, mince pies and a savoury of anchovy and cod's roe. 'Unstinted' champagne and liqueurs. Table strewn with dishes of burnt almonds, crystallized fruits, chocolates 'and such toothsome kickshaws'.)

Opposite, top: *Captain Scott's main hut at Cape Evans*

Bottom: *Captain Scott's bed. (Partially dissected emperor penguin on the table)*

I was touched to see, from the plan in the hut, that they had a pianola. (It's gone, now.) I wonder what rolls they played? I also came upon a copy of *The Strand* magazine for March 1907: W. W. Jacobs, Warwick Deeping, Max Pemberton, E. Nesbit.

This hut is roofed with four transoms. It has only three windows, each about 3 ft 6 ins square, so that, like the other at Hut Point, it's dark inside until your eyes get used to it.

As in the smaller hut, there was a great deal of equipment everywhere, from tin mugs and flat irons to a hockey stick, skis and the circular cane snow-shoes for the ponies (which they wouldn't wear – see Scott's entry in his journal for 29 November 1911). The big, twelve-foot pony sledge was standing upright by itself, not far from the entrance.

Peter was deeply moved. We all were, of course, but he in particular. As others went clicking and chattering about, he stood aside and waited silently. At last, when everyone but about five or six English – including Tony Irwin, Alan Gurney, Ronald and I – had left, he set methodically about photographing everything – almost every individual item in the hut – with the greatest care and attention. It took him a long time.

During the afternoon the ship moved a few miles further north to Cape Royds, where Shackleton built his hut in 1907. Using this hut as a base, he and three companions got within about a hundred miles of the South Pole in January 1909. (He was also the first man to climb Mount Erebus.) On the way to Shackleton's hut we walked through a herd of Weddell seals. The hut itself was surrounded by penguins who had, as Ronald remarked, a proprietary air. (Even Ronald can't help anthropomorphizing penguins.)

About six o'clock we set sail westwards for Cape Hallett. The lounge was rather quiet that evening. I think a lot of us felt preoccupied by thoughts of Captain Scott and his party.

Shackleton's hut, with attendant penguins

Inside Shackleton's hut

8

MACQUARIE ISLAND

As we left Ross Island the weather, which up till now had been almost phenomenally good, took a swift turn for the worse. The wind – a following easterly – grew very strong, with driving snow which froze to the deck, while fog to port obscured the wild, precipitous coast along which we were sailing. I found it fascinating to stand on the bridge – everyone was free to go on to the bridge at any time, ask questions and study the charts and instruments; there never was a more easy-going ship – and peer into the radar screen for the outlines of icebergs and coasts hidden in the fog. Prospero himself couldn't have been better informed about things in proximity but out of sight. Better still was to watch the antarctic petrels, which seem to delight in gliding and playing round ships. They would hover, with quick-beating wings, for all the world like English kestrels; then suddenly turn out of the horizontal into a vertical plane and, with their pointed, brown-and-white wings outstretched, sweep down to sea-level, up again and past the bridge, across the bows and out to sea – all in a few seconds. Half a minute later the same bird would return and repeat the trick with variations. These petrels are complete masters of the harshest environment in the world; or certainly the harshest of any inhabited by warm-blooded creatures. They don't just endure it; they're at home in it.

The captain had intended that we should land on Cape Hallett, where there is an emperor penguin rookery. The breeding season was practically over, but we were hoping that perhaps a few of the birds might not yet have dispersed to sea. However, we were unlucky. There could be no question of launching zodiacs in such a sea as this; so on we pressed for Cape Adare.

It is a pity that we saw no emperor penguins, either at Cape Hallett or anywhere else throughout our voyage. We had just missed them. They had not yet returned from sea to begin their extraordinary winter breeding cycle.

The emperor penguin is a miracle of antarctic evolution. Not less than 150,000 pairs breed in about twenty-three colonies scattered along the icefoot of Antarctica. After a summer of heavy feeding on krill, squid and small fish, the adults are very fat when they come ashore, in late March and early April, over the new-frozen inshore sea, at the season when all other visible life is deserting Antarctica.

With unerring sense of direction, they march south in single file to reach remembered nursery sites far from the sea. These are always on solid ice and generally close to glacial walls or high pressure-ridges which give protection from the build-up of winter snow. For about six weeks the males shuffle around on the ice and find partners in squawking courtship of the females.

Marking of these penguins with identity tags has proved that in nearly all cases partners remain together for life. Partner-recognition is by voice.

On the frozen breeding-grounds no food is accessible. The sea steadily retreats – perhaps for as much as 125 miles – as the winter ice extends outwards from Antarctica.

When at last, in May, the female lays the single large (about 0.5 kg) egg, there is much excitement and mutual 'talk'. The male awaits its appearance intently, and with his curved beak at once rolls it over his feet and up into a kind of pouch between his legs, where it is protected by a large flap of feathered belly skin and warmed by contact with the naked, hidden brood patch. If he did not do this, the egg would freeze within one minute. Exhausted by her efforts, and starving, having lost much weight during the long fast of mating and egg-building, the female now waddles seaward, tobogganing down slopes and now and then sleeping for short periods among the ice-hills.

As egg-laying in a colony is almost simultaneous, the females join to form marching lines, moving seawards with a swaying gait and huddling together in any blizzard which comes down on them. To reach open water means a journey lasting many days, even weeks. The females, however, are in no hurry to come back. They must feed well to restore their body fat in preparation for the march back to the rookery, to which they will not return for two months.

Meanwhile the mate is fiercely possessive of the egg within his pouch. Should it by accident roll to the ice during preening or testudo-forming movements, he hastily repossesses it. He must be quick, not only because of frost, but also because neighbouring males – the unmated, or those who have lost their eggs through some misfortune – will eagerly try to pirate it and beak it into their empty belly pouches. So keen is this male drive to incubate that they will sometimes 'steal' a stone or a round lump of ice; and later on in the breeding cycle they will, if they get a chance, appropriate a live or dead chick. Frozen eggs, lost during testudo-forming, they will incubate until they become rotten and burst.

Incubation lasts two months, throughout the coldest period of midwinter darkness and frequent blizzards, during which the males survive only by crowding together in a solid shield, or testudo, to conserve warmth and avoid being blown to destruction. A testudo may be completely snowed-in, but if it is, the snow forms a warm blanket protecting the outer ring of individuals, while the combined heat of the bodies melts the top snow, preventing suffocation. The testudo may sink a few centimetres during a blizzard of long duration, as the birds' feet melt the 'standing-room-only' floor of ice. In calm weather the birds shuffle apart, taking short, mincing steps to ensure that the egg remains safe in the pouch. A fresh testudo is formed in the next blizzard, not necessarily on the site of the old one, but depending on the direction of the wind.

The rookery remains clean, for since the males have eaten nothing they void no faeces. They relieve thirst by taking an occasional mouthful of snow, and gradually lose half their prime weight.

Surely there can be nothing left in that emaciated stomach to feed the hatchling when at last it bursts forth from the pouch? The male responds with tender groans to the chick's squeaking struggle to be born, his bill ready to push it back if it should expose its wet skin to the frost. Normally the heat of the male's brood-patch dries the chick as it emerges from the egg. It remains warmly hidden for a day or two, not requiring to be fed until it has completely absorbed the egg-yolk with which it is born. The empty egg-shell, however, falls to the ice, and may be nibbled and eaten by the parent – a rather inadequate interruption of his three months' fast.

The male's stoic, heroic devotion to his duty as incubator and nurse must be unique in nature, involving that almost incredibly long fast under conditions of exposure to intense frost that would kill most other living creatures. It is at last rewarded, while the rookery is still sunless in July, by the return of his mate, fat and full-bellied from her long sojourn amid the krill and small fishes. She has had an even longer walk back to the rookery, since winter ice is still forming far at sea. She usually arrives a few days after the chick is born at the time when, getting hungry, it begins to poke its head into the air and whine for food. The male, by an unusual provision of nature, manufactures sufficient nourishing fluid from bile and stomach secretions to keep the infant alive until the female arrives.

When the female returns, she calls to and recognizes her mate by voice. This is a kind of ceremony, which may take some time, since after two months of testudo and other movement the mate is not likely to be where she left him nursing the precious egg. Once the ceremony of vocal recognition is over, the female persuades her mate to yield the chick to her. Within seconds it is transferred to her pouch. The male, in his turn, is now free to set out on the long walk to the ocean feeding-ground. Off he goes, and won't be back for at least a month or more. He needs all that time to restore lost condition and then walk back again. And here we find another remarkable and unusual natural provision: the mother is able not only to live off her body fat but also to conserve the contents of her stomach to dole out enough daily food to keep the chick going until the male returns. By some device which we do not understand, she is able to retain food in her stomach for several weeks without assimilating it.

Now we begin to see the main advantage of winter breeding. By the time the chicks are old enough to leave the pouch and huddle for warmth in the groups (called 'crèches') which they form during September, the sun has returned and the distant sea-ice is breaking up. Both parents are thus free to go fishing, making progressively shorter journeys to obtain larger, more frequent meals for their chick. The chicks recognize the parent birds, returning with food, entirely by their individual voices.

At six months the emperor youngster – an engaging creature, with black nose and eyes in a facial rosette of white – sets off with its parents for the sea. It is now very fat, though the mother and father are lean. The family embark on an ice-floe under the

twenty-four-hour midsummer solstice sunlight, and are borne
away on the ocean current. The young birds 'porpoise' and play
fishing games in the water, while the old birds put on enough fat
to enable them to rest and sleep away the annual moult.

We reached Cape Adare at two o'clock one afternoon, in a
Force B southerly wind. I have never seen any place more forbidding and bleak, or any sea breaking so wildly upon a more
desolate shore. We might have been the only people alive on
another planet – some uninhabited and inhospitable world.

For the next twenty-four hours, as we held on our way north-
ward, we tumbled and pitched like a dead cat in a weir. The
Lindblad is not a large ship. Her gross tonnage is 2,345 and her
overall length 239 feet. We were rolling 20° to 25° on average,
but now and then would come rolls of 35° and 40°. At dinner the
entire contents of our table, seating six, shot into laps and over
the floor. I had never seen this actually happen before, and I
doubt it's of much interest to anyone who merely reads of it; but
I don't think any of those who were sitting at the table will forget
it in a hurry. The Chinese stewards took the incident in their
stride with outward impassivity, simply picking everything up
and re-laying the table, notwithstanding that the same thing
might easily have happened again at any moment. We settled for a
rather more exiguous spread and held on to everything by hand.

> When the cabin portholes are dark and green
> Because of the seas outside;
> When the ship goes *wop* (with a wiggle between)
> And the steward falls into the soup-tureen
> And the trunks begin to slide;
> When Nursey lies on the floor in a heap,
> And Mummy tells you to let her sleep,
> And you aren't waked or washed or dressed,
> Why, then you will know (if you haven't guessed)
> You're 'Fifty North and Forty West!'

(We were actually Seventy South and a Hundred and Seventy
East.)

The heavy seas continued. I don't think anybody slept much
during the night; not through fear – we all felt secure enough –

but simply because we were all waving about in our beds like washing on a line. Ronald, whose bunk lay fore and aft, had to strap himself in. I, lying port and starboard, was less in danger of rolling on the floor. Although I lay awake for four or five hours, I found I felt perfectly relaxed and content. I wonder why? No worries, I suppose, and the knowledge that there was nothing much that had to be done on the morrow.

After breakfast, while we were still rolling strongly, Ronald and I repaired to the bridge to see whatever there was to be seen. The ship was proceeding with due care and attention, for the sea all round us was full of 'growlers', which needed watching. A growler is an almost-submerged piece of ice, bigger than brash, floating in a seaway. They vary in size, of course: anything from the size of a bed to the size of a good, big field. With the sea washing over them, they are often only intermittently visible; and since they may not always show up on a radar screen as distinct objects, it's desirable to keep a weather eye open. Nowadays their 'growling' (*vide* 'The Ancient Mariner') is not always audible above the sound of a ship's engines, but in the days of sail it was a noise only too well-known to antarctic sailors. Q: When is a growler not a growler? A: When it is a

Moraine near Cape Adare

bergie-bit. A bergie-bit is a lump of half-submerged ice, bigger than brash, which is *not* floating in a seaway, and does not therefore merit the more sinister name of 'growler'.

As compensation for all this rocking in the cradle of the deep, there was any number of birds to be seen: antarctic fulmars, very much like our North Atlantic fulmars in behaviour and appearance – that is, fluttering and gliding, and with the same softly rounded, white faces – but the plumage paler, the wings whitish-blue and flecked with brown on the primaries; mottled petrels, their heads a kind of white, splashed with brown, the wings dappled brown on white; and black-browed albatrosses, never ceasing to seem marvellous in their swooping flight, gliding beside or aft of the ship on great pointed, backswept wings, black above, but below white-centred and edged with black (like visiting-cards used by people in mourning, long ago in the 'twenties and 'thirties). As already described, the tilted, black slash of the supercilium ('eyebrow') gives them a scowling, tough expression at close quarters, and I dare say this is one bit of anthropomorphism which may not be altogether inappropriate. Old Betty Robinson stood feeding them plum-cake from the stern. She wasn't much bigger than an albatross herself.

Nunatac (with Catabatic wind)

About half-past four that afternoon, as we were passing Sturge Island, a great flock of sooty shearwaters flew across the bows. There must have been fifty or sixty of them – smoky dark, but smudgey white on the undersides of the wings – flying in the way characteristic of all shearwaters, dropping sideways at right angles, one wing down, almost to the surface and then gliding up again. Ronald, expatiating at my request, explained that, in their falling flight, they would actually strike the water if it were not for the cushion of air compressed between the sea and the bent tip of the wing. 'One of the most numerous birds in the world, the sooty shearwater,' he added.

One day we were all asked, as part of some research project on oceanic drift, to write our names on labels bearing the name of the ship, the date, the day's latitude and longitude and a return address, cork them into empty bottles (one per person) and throw them over the stern. It wasn't merely sport, for ships, of course, don't often get into these remote parts and nobody really knew where our bottles might reach land – not even Dr Ted Walker, the official oceanographic lecturer on board. There was something about this exercise which seemed to please every-one and put them in good spirits. Betty, Miss Hartwell, Bob Leslie, Carolahn Best and all the rest of us threw our bottles astern with a will and watched them receding, fragments of identity, very small and fragile-looking, down the *Lindblad*'s white wake.

So after a few days of boisterous sailing, during which the weather grew warmer and warmer until the thermometer on the bridge stood at $+7°C$, we found ourselves, one evening, off Macquarie Island, latitude 55° South, longitude 159° West, about 600 miles south of the southern tip of New Zealand.

Macquarie is a dependency of Tasmania. It is a long, thin island; to be precise, twenty-one miles long and anything from a mile and three quarters to three and a half miles wide. There's a central plateau about a thousand feet above sea-level. At the northern end a narrow, low-lying isthmus connects the main part of the island with the northernmost tip, known as North Head. The whole island is fringed with rocks and reefs, so that landing is not terribly easy, more especially on account of the great masses of heavy kelp in the sea all round the shores. An odd feature of the place is the large number of small lakes scattered all over the upland plateau.

In 1896 a certain Joseph R. Burton, taxidermist to the Colonial Museum of Wellington, New Zealand, landed on Macquarie. (He was not only a taxidermist, as will be seen later.) He remained there until 1900, and had this to say: –

It is an exceedingly rough place, consisting simply of a series of almost bare hills, the highest reaching an altitude of 1,300 feet above sea level. Macquarie Island is evidently of volcanic origin, judging from the lakes on the top of the hills, filling ancient volcanic vents. The prevailing winds and storms are westerly. Clouds, as a rule, envelop the hills; therefore, the region is always damp, while the rainfall is probably four times greater than the average, say, of that of New Zealand. Streams, not to mention bogs, springs and swamps, are plentiful everywhere, except, perhaps, at the north end. Some of the upland lakes are curiously margined with moss, which grows so thickly as to extend over the surface of the water towards the centre of the lake. Such places are pitfalls. When you fancy you are walking on the solid shore of a lake, your legs suddenly pierce the vegetable crust, and you find yourself up to the armpits struggling in water of doubtful depth. In such a place I endeavoured once to find bottom with a ten-foot pole and failed. In one lake there is a floating island of moss several acres in extent.

There are no trees, or bushes even, of any kind on the island, the chief vegetation being rank and tall, yellowish tussock grass, the bright green Macquarie cabbage, so called, and the large cushion-shaped masses of azorella and sage-green, close-growing pleurophyllum of the botanists. Of course, there are other minor plants, including one or two ferns, while mosses grow abundantly in places. When the moss beds become frozen you can walk over the surface without crushing them.

We arrived in fairly dense fog and lay to all night, much looking forward to going ashore next morning. Apart from anything else, we had now been at sea (and rough at that) for a good many days past.

That night in the lounge, two or three of the professional naturalists on board, including Dennis Puleston from California, were amusing themselves (and me, I may add) by inventing a ribald ornithology.

'The other day,' remarked Dennis, 'on the after-deck, I observed a fine specimen of the Rosy-breasted Pushover.'

'The bird is much harassed, I believe,' said someone else, 'by a predatory species – the Gimlet-eyed Titwatcher.'

Various other birds, it appeared, had been observed on board: the Furtive Nutscratcher, the Horizontal Bedthrasher, the Extra-marital Lark. And one man – could it have been Tony Irwin? – had more than once seen, in his own home, the Exorbitant Gasbill.

Later, as I sat having a drink with Tom Richie, a young naturalist and wildlife painter from Florida, he told me he could sign his name in Chinese. 'I'd like to see that,' I said; and accordingly Tom executed the following on a page from a notebook:

李志德

I asked the Chinese barman, Honghi, to be good enough to come over and read it. Honghi read off 'Lichee Duh'.

'What's Duh, Tom?'

'Chinese approximation to "Tom".' (In Chinese, of course, the surname precedes the 'given' names.)

Seven o'clock in the morning and woken by Big Dave over the ship's intercom. 'Before going up to North Point we're taking the ship inshore towards the king penguin colonies along the east coast. If you want to see the king penguins you should come on deck right away.'

Worse than Aldershot in the war! Cursing up out of sleep, dragging on shirt and trousers over pyjamas, grabbing field-glasses, scarf, where's my specs?, oh hell! Ronald and I staggered along the corridor, meeting Peter, whose artificial tooth had fallen out: he was miserably clutching it in one hand. We emerged on deck into the usual bitter wind.

And then suddenly – in a moment, in the twinkling of an eye – everything took on a different aspect. We were about half a mile off-shore and moving gently landward. The coast of Macquarie reminded me of nothing so much as that of Sierra Leone. Steep, vivid green, vegetation-clad hills rose from the shore to be lost in swirling mist at a height of five or six hundred feet. Here, however, the resemblance ceased, for this place was bleak and desolate. Here and there were bare slopes, each a few acres in extent, and these, seen from this distance, appeared oddly speckled in places, as though with white flints.

Approaching Macquarie. Penguins (kings and royals)
seen from off-shore

The open beaches, too, were white in colour. It took us a few moments to grasp the reason. The entire shore of the island was thick with king penguins, standing almost shoulder to shoulder. I have never seen such a vast host of birds and it's safe to say I never shall again. They were numerous beyond belief, millions of birds, covering the open areas almost like swarms of bees.

As we stood gazing at them incredulously, by degrees coming to terms with this amazing sight, we began to become aware also of our closer surroundings. Overhead were flying any number of birds – a few skuas, a light-mantled sooty and a black-browed albatross or two (by now we had become quite blasé about black-browed albatrosses), several black-and-white antarctic cormorants (*Phalacrocorax varius*, not to be confused with the blue-eyed shags we had seen earlier in the trip) and a great many of our old friends the giant petrels; mostly black, but a few pure white – the same bird, for here the white strain, which for the most part is found further south, occurs side by side with the more northerly dark variety.

All round the ship king penguins were swimming and 'porpoising'. Our propellers, very gently though they were set, in order not to hurt the birds, had churned up the krill, and this was the attraction. The water was actually thick with king penguins. They were like running salmon above Galway Bridge, for they were so close together that you might have supposed you could walk on them. Their cry is startlingly like a human voice: 'Ho!' 'Ho!' 'Illo!' 'Illo!' (The above-mentioned Mr Burton, in his account of Macquarie, describes the cry as sounding like 'I have not got time', but I couldn't quite hear that.) Standing in the centre of the ship, however, and not looking out to sea, you would have thought that these were human cries. I'd like to have made a tape-recording.

These king penguins are very beautiful – far more striking and beautiful than the smaller gentoos, Adélies, chinstraps and rock-hoppers. They are big – about two and a half to three feet in height – and have long, prominent, relatively slender bills. That is to say, the bills look slender in relation to the bulk of the bird, and by contrast with the thick, squat beaks of smaller penguins. The neck, too, is slender and long in appearance. There are no less than six vivid colours (counting black and white) to be seen on a king penguin, and five of them come together most arrestingly in the area of the throat. The head and face are deep black and the chest and belly white. The back, however, is a sleek, speckled grey, and this grey comes forward almost to meet round the throat, as though the bird were wearing a grey cape clasped at the neck. At the back of the head, one on either side, are two extraordinarily vivid, deep-yellow, velvety patches, each about as big as the palm of a human hand. They are shaped like commas, the tails of which curve downward and forward towards the throat, ending about the shoulders of the 'cape'. Just below the throat, at the front, is another band or patch of this rich yellow, which shades downward to pale lemon blending into the upper part of the white chest. The whole effect is to give the bird a flamboyant, dandified appearance – rather like Oscar Wilde off to a dinner-party. (And they *are* rather fat, too.) The sixth colour is found in the lower mandible of the bill, which is a smooth, matt russet-orange.

Merely to stand and look at this sea, turbulent with penguins and roofed with circling petrels and cormorants, aroused everybody. People were catching each other by the sleeve – 'Oh, look there! Look!' – and running from side to side like children. Betty Robinson had tears in her eyes. 'Did you ever see anything more beautiful, dear?' 'Watch out, Betty,' said somebody, 'they're looking for recruits.' In our high spirits it seemed the best of jokes and Betty pretended to start climbing over the side.

The speckled slopes high up on the coast, we now realized, were colonies of royal penguins. The 'royal' is a good deal smaller than the 'king' and tends to go up the cliffs to breed because of the domination of the 'king' on the open beaches.

As already described, the largest and most antarctic of all penguins, the emperor, has an annual breeding cycle which ensures that its chick reaches independence at the most suitable moment of the year – just after midsummer. Its mating and incubation take place in midwinter. The smaller king penguin, despite breeding in warmer, sub-antarctic shores, manages in a very different way the same problem of raising a large, slow-growing chick in a climate where for half a year food is hard to find. The colony is occupied throughout the year by birds in different stages of growth or moult. The single egg, laid in November, is incubated, like that of the emperor, in a brood-patch covered by a flap of belly skin. (It was an old joke aboard the early sealing ships that you had only to lift up a king penguin by the scruff of the neck and shake it to make it lay a fresh egg.) No nest is made; the bird shuffles around with the egg on top of its feet and warmly hidden under the flap.

The chicks we saw on Macquarie were mostly just hatched from the November layings. They were growing fast, but not all of them fast enough to be ready to take to the sea before the plankton and krill sank for the winter in April and May. At this time the parents find it hard to get enough food even for themselves. They do not altogether abandon their half-grown children, but they come to feed them only at long intervals of sometimes more than a month. The weaker chicks die, but the larger ones huddle together for warmth in crèches. They lose weight through the winter. Then, in October, the krill rise to the surface again – it is early spring – and the parents return more often to feed the chicks. The chicks reach mature weight, and in late summer are moulted into their first full adult plumage. The spring feeding of the chicks means that the parents are too much preoccupied to be able to breed in that same year.

However, there are always some adults which have lost eggs or chicks early in the season, and these will be the first to return and breed next November. According to Bernard Stonehouse, well-known as an authority on penguins, the successful early breeders of the previous season often become the late breeders of the current season. Such late breeding, however, is usually a failure. 'King penguins stand only a sporting chance of raising two chicks in each period of three years, or of one every second year if their late breeding fails.'

It was clear that landing was going to be tricky. Inshore, the island is thick with *Duvillea antarctica*, a sea-kelp which is the *longest* plant in the world, and one of the toughest. Its fronds often stretch several hundred feet from its sucker or 'anchor' on a rock or stone. Also, the waves were breaking quite heavily on the stony beaches and one way and another it didn't look too promising.

The ship proceeded towards the north end of the island and the narrow, flat isthmus, which is almost level with the sea. Here is situated a Tasmanian meteorological and observation station, and beyond lies the North Point. We gave the point a wide berth, for here great, jagged rocks extend a long way out to sea. At length we turned about, through a hundred and eighty degrees, came back down the west side of the point and anchored in Hasselborough Bay. Here we lay to while some of Big Dave's merry men went out in a couple of zodiacs to make a reconnaissance. They reported that landing was feasible for those who didn't mind a rough ride and a possible wetting on landing. The more elderly, etc., were advised to wait a bit.

Well, it certainly *was* a bit of a rough ride. Sitting in the plunging zodiac, while Filipino sailors, as best as they could, held it with ropes close under the gangway stair and bundled the stumbling passengers aboard one by one, I began to wonder whether perhaps I'd been foolish. The Swedish doctor's wife, Britt; the beautiful lady purser; Mr Bob Leslie, Miss Carolahn Best and various other spectators were looking down from the main deck (how far above they seemed!) with expressions of rather ill-concealed anxiety, I thought. Lord, how we bobbed up and down! dropping six feet or so below the lowest step of the gang-plank, then surging up above it on the next wave. People were pitching themselves down into the zodiac just as opportunity offered, and then – since I had been put to sit amidships, directly opposite the boarding-point – falling all over *me*. However, get away we eventually did, and it was a bit less alarming once we no longer had the side of the ship towering over us. We bumped and thudded the best part of a mile, coming in among the dark,

Duvillea antarctica *at Macquarie Island*

smooth, rubbery, tossing ropes of *Duvillea* – which have to be seen to be believed – and grounding on a rather steep beach, where the waves were coming and going with a lot of animation. Here Keith Shackleton, Jim Snyder and other stalwarts in waders were standing in the surf to hold the zodiacs steady while we clambered over the side – this was the wetting – and stumbled to shore. My moon-boots did me yeoman service and I didn't ship a drop.

The young fellows of the Tasmanian station made us welcome and treated us as hospitably as ever the monks in the Cévennes treated Robert Louis Stevenson and his donkey. (They had some very good beer, too, which they told us they brewed themselves. It was much the best we'd had since leaving England.) Unexpectedly, their resident doctor was a girl – the wife of one of the meteorologists – and this lass offered to take a few of us up a steep gully, some way off along the rocky shore, to see two light-mantled sooty albatross chicks in their nests. It would be a bit of a climb, she said, but worth it.

Skua devouring elephant seal carcass

On the way we passed three skuas tussling and ripping over a very dead elephant seal – a nasty sight. I hadn't seen skuas so close before: I was within a few feet of them. They were rather bigger than chickens, with frightful beaks.

As for the elephant seals – which were all over the beach – they were colossal; each weighing, I was told, a couple of tons. Obscenely fat, they lay moulting in lethargic groups, taking no notice of us.

In the spring – which is September – the big master males of the elephant seal (Mirounga leonina) come ashore and set up territories – for which they fight other bulls. A fortnight later the gravid females arrive and select which 'harem' they will join. Here, in October, they give birth to their single pups, which weigh a mere 100 lbs. These grow at an amazing rate, fed on their mothers' extremely rich milk, so that in less than four weeks they are weaned. The mother's milk has now dried up, and she has mated several times.

The elephant seal is the largest seal in the world, weighing up to 8,000 lbs, or over seven tons. The mature female is only a quarter the weight of a beachmaster bull. The new-born pups are attractive creatures, with woolly black coats. They play together in the shallows and kelp while their parents, who have eaten nothing for a couple of months, go off to feed and fatten in the deep sea (no one is quite sure where). The youngsters learn to fend for themselves, and as soon as they have lost their juvenile black coats and most of their puppy fat, they too depart to feed in the ocean. They don't come back for quite some time. Meanwhile the adults, fattened up again, come ashore to laze in the midsummer sun. They moult their worn-out fur and the skin with it. At this time they do not mate. The cows are already pregnant, though the embryo does not develop – it goes through a latent stage – until the moult is over. At this season the bulls never fight. All they want is to sleep and to get rid of their old fur.

After leaving the beach and turning inland, we had a clamber-ing scramble up the steep, stony bed of a little stream cascading down from the hills. I wished I'd had my fell-boots instead of the soft, spongey moon-boots, which gave ankles less protection than even gumboots would have done. However, as the girl doctor had promised, the climb *was* well worth while.

The two nests were sited on a grassy ledge near the top of the gully, easily accessible, and not more than a few feet apart. They were simply thick, hollow circles of earth, like children's sand-castles, about eight inches high and perhaps eighteen in diameter. In each sat a chick.

The sooty albatross lays only one egg each year. As the chick gets to about six to eight weeks old its appetite becomes enor-mous, and both parents spend most of their time foraging, and filling their crops to feed it. This morning all four parent birds were away from these nests, no doubt getting krilled up for the next regurgitation. Like almost everything else in the Antarctic, the chicks showed no fear of us and made no show of ejecting throat-fluid as we stood round them (photographers clicking away for dear life). They were big – bigger than ravens – but a lot of this was the fluffy, grey down with which albatross chicks are covered until they grow their first-year feathers. This down was so light that it moved continually, like cobwebs or gossamer, in the almost still air. Their great, black beaks, like pickaxe-heads, contrasted oddly with their immaturity. They watched us alertly – a little tensely, I thought – but we kept a few feet away from them and they evidently felt no need to cry out or leave their nests.

Returning at length to the beach, we found the devoted staff of the *Lindblad* once more at their task – this time of getting the beached zodiacs launched for return to the ship. The waves were, if anything, more troublesome than before. A zodiac is normally launched from a beach by pushing it out until it floats freely and then holding it more or less steady on ropes while the passengers wade and scramble in over the inflated, convex sides. Just now, however, this method, despite all the efforts of Big Dave and his henchmen, was proving impracticable, for every time the zodiac was pushed out the waves flung it back and grounded it. Finally the exasperated Dave was driven to extreme measures. 'Get in there, Snyder, and get your arse wet!' Thus adjured, Mr Snyder (at risk, I would have thought, but without the least hesitation) waded into the rough sea, dragging the zodiac behind him by a rope until it floated, and then held it on the rope by sheer physical strength, while the waves did their

best to drive it back on shore. He continued to do this as we
embarked. Fortunately he didn't get knocked over, as I doubt he
could have got up again. This sort of thing was all in the day's
work to Dave and Co.

Next morning Ronald and I joined others in a nice, leisurely
walk – about three miles – along the rocky eastern shore of
the island to a beach known as The Nuggets (from some conspicuous rocks nearby). Here there were big colonies of both
king and royal penguins.

The royal penguins turned out to be quite delightful – for my
money the most attractive of the twelve species of penguin we
encountered altogether (magellanic, gentoo, Adélie, chinstrap,
erect-crested, rockhopper, macaroni, king, royal, Snares,
yellow-eyed and blue). They are much smaller than kings –
about eighteen inches tall. The stout, thick, almost puffin-like
bill is a smooth, dull orange all over, and on the head is a yellow,
backswept plume – a kind of *panache*, neatly parted on either
side, so that it looks rather as though the bird had arranged it
that way. Royals are white from chin to foot, with black backs.
The eye is the same curiously dull, 'wall' russet as the king
penguin's. The cry is a long, harsh caw, followed by a loud
rattle.

Ronald told me that royals lay two eggs, one of which they
simply ignore. 'Why?' 'Well, some time during the last million
years they've discovered that one egg's enough to maintain the
species. In another million years they'll probably be laying only
one.'

These royals were the tamest of all the penguins we met. All
the other species, while unafraid of us, kept a little distance – a
yard or two – and walked away if you came any nearer. Not so
the royals: you could stroke them. I picked one up bodily, as one
might an infant. It showed no fear at all, but remained quite
placidly in my arms until I put it down again. (It could have
pecked at my eyes if it had had a mind to.)

After our own clumsy troubles in landing the day before, it
was amusing to see the penguins 'porpoising' in through the
heavy surf, coming to land in shallow water, standing up and
waddling ashore as though there were nothing to it at all – as
indeed there wasn't, for them.

After the three-mile walk along the beach I felt thirsty. There
was a little brook running out over the beach. 'I suppose one can

Royal penguins
(Eudyptes schlegeli)
'porpoising' in to
Macquarie beach

drink this?' I remarked to Ronald. 'It doesn't look any different from a beck in the Lakes.' 'Well,' he replied, 'there's a huge rookery of royals inland, about half a mile upstream, so it's probably about forty per cent penguin piss, but you could try.' I decided against it.

We made our way up the course of the brook as far as the rookery. This turned out to be a flat area of about two acres or thereabouts, which over the years the birds had trodden bare. They were very numerous and included a lot of well-grown fledglings which had partly lost their grey fluff and gone into adult plumage. The inevitable skuas were flying back and forth overhead.

It was here that a rather amusing little incident took place, involving a white giant petrel. These big, heavy birds can't take off in flight unless they can first run for a little distance, flapping their wings. This one had got itself into the bed of the brook. Upstream was a lady member of our party, while downstream was myself. Behind the bird was a sheer bank a few feet high, topped with great tussocks of thick, coarse grass extending

Fur seal
(Arctocephalus
australis)

inland. In front were the shoulder-to-shoulder crowds of count-less royal penguins, two hundred yards deep. The petrel, though no one was molesting it, was obviously anxious to depart. It didn't fancy passing either of the humans, and it saw no future at all in the bank and the deep, rough jungle of herbage beyond. Finally it decided to have a go at getting through the penguins, and accordingly hurled itself at them. Giant petrels have formid-able beaks and are fairly massive. The penguins, however, weren't giving way at any price. Closing ranks tightly, they cursed and swore at the petrel. Its sheer weight enabled it to penetrate a yard or two, but of course this only brought it deeper into the mass of the penguins, which turned inward and mobbed it from all sides. It got itself clear and retreated into the stream, but a minute or two later tried again. Same result. Finally it thought better of the whole thing, scrambled up the bank and disappeared, struggling away through the long grass.

In the afternoon we were taken by our hosts of the scientific station to see some rockhopper penguins on another part of the shore. They were not on a true beach, but a rather steep, uneven terrain of fallen rocks and boulders – a somewhat inhospitable nesting area, I would have thought. They were not altogether dissimilar from the royals, but less striking and attractive. They, too, have white fronts, black backs and a yellow plume or tuft on each side of the head, and they also have a very distinctive, shrill cry, 'Kille*rik*! Kille*rik*!'

Near by were a few fur seals – most beautiful creatures. After the lethargic, clumsy elephant seals sprawling all along the shores of the island, these seemed elegantly lissom and graceful. They were each about seven feet long, with very fine, soft fur, and when not recumbent sat up alertly, on four flippers, holding the fore-part of their bodies erect as they looked here and there about them.

Coming back we ran into a couple of *wekas* (a Maori word, pronounced 'wecker'). The weka is a flightless rail, and a most predatory, aggressive bird. It is native to New Zealand and was originally introduced to lonely Macquarie (and to other southern islands) by the sealers, for food. It is something larger than a moorhen, of a russet appearance, speckled black and brown, with a sharp bill and an inquisitive, questing manner. It wanders about in a moorhen-like way, on long, orange-coloured, back-bent legs, pausing at each speculative step. It can kill (and eat) rats, and will, so Ronald told me, come fearlessly and impudently into a camp and, magpie-like, take watches, rings, bootlaces and any other little objects that catch its fancy. These two certainly showed no fear of us.

I shall end this chapter with a short account of the activities of Mr Joseph Hatch and his colleagues and employees on Macquarie Island between about 1890 and 1918, since the said Mr Hatch well exemplifies what human beings are prepared to do to wildlife for commercial gain. And if you think that Mr Hatch is a thing of the past and couldn't happen again, just reflect upon the African ivory trade and the Canadian harp seal slaughter, both still going strong.

Joseph Hatch was born in Australia in 1837. Round about 1890 he began his exploitation of the wildlife of Macquarie Island, which belonged, then as now, to Tasmania. Hatch and two men called Hawke and Henderson were directors of the Southern Isles Exploitation Company Ltd, which invited investment and issued stock in the normal way. (I've seen a copy of the prospectus.) The work of the company's employees consisted of expeditions to Macquarie, where they slaughtered penguins (kings and royals) and elephant seals, and boiled the carcasses (or in the case of the seals, selected parts of the carcasses) for oil. The oil so obtained 'proved of good quality and could be utilized for soap-making and leather dressing'. It was barrelled and shipped back to Tasmania, New Zealand or Australia as convenient.

The above-quoted Mr Joseph Burton spent his three and a half years on Macquarie (from 1896 to 1900) as a 'collector' for Mr Hatch. His account includes the following:

> The royals are good for the amount of oil they yield when boiled down. Boiling down takes place from January to March. As many as 2,000 birds can be put through the digesters in a day, equal to fourteen casks of oil, each about forty gallons.
>
> By placing a fence across their path the penguins, when coming from sea, can be easily driven and yarded like sheep. When the yard is full, ten men go out and club the birds before breakfast. When work is resumed many of the poor birds are found to have recovered and are walking about; they require re-clubbing. The bodies, as they are, are passed into digesters, and boiled by steam for twelve hours. The crude oil is then blown into coolers, casked, and is ready for removal to New Zealand, or elsewhere, for refining.

On the beach at The Nuggets there remain at the present day four or five great, cylindrical, rusty cauldrons (and I mean 'great': I haven't got the measurements, but they tower over your head and they're very broad in diameter) of cast iron, and reminiscent, with their heavy, screw-down lids and drain-off taps at the bottoms, of the old-fashioned wash-house 'copper'; together with the remains of other, ancillary equipment. These are what the company described in an inventory as 'digesters and steam boilers'. (There are five more at various other places on the island.) It was in these that the bodies of the penguins, totalling hundreds of thousands, were boiled in steam for their oil. Hatch himself described the operation as follows:

> There is a platform at the top on to which the birds are thrown, so as to be put into the open space at the top of the digester, for there is a man inside to stow the birds as if putting in bricks. The birds are killed with a staff with a ferrule at the end and die immediately they are struck on the head. [As we have seen, Mr Burton knew otherwise.] The young birds, one year old, which we kill, are known by having

only one feather on their heads at that time ... The birds are killed to-day to load the digester to-morrow, and are very often cut open and cleaned to allow the oil to be extracted quicker by boiling in the small digesters.

Hatch and Co. continued their profitable activities for the best part of thirty years. (I have seen a photograph of the elderly, spectacled Hatch. It reminded me of Sir Bernard Miles got up to play the part of Joe Gargery.) However, early in 1918 trouble began. It was originated by none other than the Australian Sir Douglas Mawson, arguably the toughest and most intrepid of all the great antarctic explorers, whose amazing exploits and incredibly courageous survival for months alone on the antarctic continent had made his name that of a hero throughout the British Empire. In January 1918, being in Europe for an international conference, he told the Agent-General for Tasmania in London that he was alarmed at the wholesale destruction of bird and animal life on Macquarie. He thought the lease to Hatch's Southern Isles Exploitation Company should be terminated and any further killing properly observed and controlled by the government. In February, the *Illustrated London News* published an article on the subject, in which it said: –

The usefulness of oil extracted from penguins was proclaimed some years ago, and with details which make all decent men shudder. The wretched birds are driven to the boiling vats up long planks, and made to precipitate themselves into the boiling water prepared to receive them. Neither soap-makers nor leather-dressers will, I take it, care to encourage a traffic so foully barbarous. But apart from its cruelty and apart from the fact that the supply would be exhausted within a few months, there are other reasons why this proposed traffic ... should be sternly forbidden.

I am not clear why the traffic was described as 'proposed', when it had been going on for years: or why, that being so, the article said that the supply would be exhausted within a few months.

The allegations of cruelty persisted, and on 17 August 1919 were given prominence by the *Sydney Morning Herald*, in an article by a man called Frank Hurley: –

The tanks in which Joseph Hatch boiled the penguins

The Penguins are mustered like sheep and driven up a narrow, netted-in runway, which terminates at the far end over the open door of a boiler digester. At this end stands a man with a club. The birds

... round a corner ... and that is the last of them; for a knock on the head and a kick send them into the boiler. It is one of the most pitiful sights I have ever witnessed. It made me feel quite sick, in fact; when the boiler is full the lid is sealed down and the steam turned on. This wanton butchery takes a toll of some 150,000 birds annually. The industry is an unessential one and, owing to the primitive plant and wastage, the profit is small and the revenue to the Tasmanian government, which leases the island, is negligible. The rookery at Lusitania Bay of the king penguins ... has been almost exterminated.

I have also seen vast numbers of sea-elephant carcasses polluting the foreshores, with the oil-yielding blubber removed only from their sides, the under-portion, owing to the trouble in handling the heavy carcase, being allowed to remain.

Hatch vigorously denied the allegations that the birds were driven into the boilers alive, as did twelve of his employees ('Well, they would, wouldn't they?' as Miss Mandy Rice-Davies would no doubt have said) in a signed document dated April 1920. The Tasmanian premier supported his denial, but could not defend his business against the heavy criticism now coming not only from Mawson, but from the Australian government, the Zoological Society of London, the Australian S.P.C.A., H. G. Wells, Apsley Cherry-Garrard (and even Arthur Mee in the *Children's Newspaper*). Mawson by-passed the hard-to-establish matter of boiling alive and other cruelty (which was never proved: the slaughter itself, of course, was indisputable), basing his case simply on the danger of extermination and the attraction of the idea of making Macquarie a wildlife sanctuary, with a Tasmanian meteorological and radio base.

'This little island,' said Mawson, 'is one of the wonder spots of the world. It is the great focus of the seal and bird life in the Australasian sub-antarctic regions, and is consequently of far greater significance and importance in the economy of that great area than its small dimensions suggest. It behoves those responsible for its good keeping to see to it that the animals resorting thereto are properly protected against any possibility of extermination ... The sea-elephant is so hunted in all its haunts that its extermination is now only a matter of a few years, unless adequate measures are adopted for its protection.' Then Mawson went on to deal with the exploitation of the penguins.

Opposite: *Campbell Island*. Top: *Hooker's sea-lion pups*. Bottom: *bull elephant seal menacing*

The penguins, having been clubbed (either stunned or dead) are packed in, the door screwed up and superheated steam turned on for

Enderby Island: below left, *saxifrage;* below right, *red-crowned parakeet.* Stewart Island: bottom, New Zealand pigeon; right, *white-faced heron in flight*

Top left: *shy albatross (left) and Buller's albatross*. Top right: *black oystercatcher*. Bottom left: *pintado on the* Lindblad's *deck*. Bottom right: *red-billed gull*

twelve hours. The fat runs away into tanks and is barrelled for ship-ment; the bones, the flesh and the feathers are dumped as refuse. It is argued that this is a very humane way of dealing with the penguins; but to those who know how tenacious of life is the penguin, a quicker and more certain method of killing is desirable if the trade is to continue ... The king penguins have been so enormously reduced by slaughter that their final extinction is threatened ... It is unreasonable to suppose that the best measures are being taken ... to preserve the island fauna for future generations.

On 2 February 1920 the Tasmanian government cancelled Hatch's licence and soon after this the Southern Isles Exploita-tion Company went into liquidation. Hatch, aged eighty-two and hotly protesting and petitioning the government to the last (on 25 March 1920, he spoke for two hours and twenty minutes at the Burns Hall, Dunedin, 'at times at such a rate as to be almost incoherent', and a motion in his favour was carried 'with acclamation'), went into retirement in Hobart, Tasmania. He died in 1928, at the age of ninety-one.

Having studied the story in detail, I am myself confirmed in the view, formed during my long opposition to the Newfound-land harp-seal killing, that when it comes to wildlife slaughter, rational arguments either way seldom go to the root of the matter. Hatred or defence of wildlife slaughter is basically a 'gut reaction'. Either people feel emotionally that such activities are vile or they do not. It is like the White Knight: –

'Everybody that hears me sing it – either it brings the *tears* into their eyes, or else –'

'Or else what?' said Alice, for the Knight had made a sudden pause.

'Or else it doesn't, you know.'

There are still plenty of people like Mr Hatch in the world. They are active, too, a lot of them. And they would move back into the sub-antarctic and the Antarctic if they could.

9

CAMPBELL ISLAND AND THE ROYAL ALBATROSSES

That night, as we were sailing away from Macquarie, there unexpectedly took place an occurrence of the Aurora Australis, the Southern Lights. I remembered the Borealis, seen from northern Ulster long ago, as a distant, shining glow filling the northern sky. This, however, bore no resemblance. The Australis looked rather like the length of a great, luminous chiffon scarf flung across the sky – a broad streamer of faint, diaphanous light, like moonlit cloud-wrack. It curved and zig-zagged from the zenith to the horizon, and after a time broke and re-formed just as cloud does. At any particular moment of viewing, however, it somehow looked more fixed and less fluid than cloud, as though made of something very light but nevertheless tangible, such as rice-paper. The phenomenon lasted for about half an hour.

Talking to one of the American naturalists on board about the small and perhaps not very beautiful mosses and small flowers we had seen on Macquarie (little stitchworts, mayweeds, chickweeds, crowfoot and the like) I asked whether one, a polystichum, had any English or local name. 'I guess not,' he answered. 'It's just a Polly Stitch'em.'

By early next morning we were sailing up into the heart of Campbell Island. Campbell is no more than about eight miles broad at the broadest points, and its principal feature is a four-mile fjord or inlet, perhaps a mile across, which carries the sea right up into the centre of the island. At the far end of this fjord lies the small New Zealand meteorological base.

Low hills, not more than about 1,600 feet, lie on either side of the inlet. It was a beautiful, late summer morning, and the place, with its bare, green slopes below craggy outcrops higher up, reminded me of the Lake District. Here and there upon the slopes we could see white dots, identifiable through binoculars as nesting royal albatrosses.

The hills were covered with mist, on which was shining a dim rainbow-nimbus from the sun astern. We anchored in Perseverance Harbour, at the top of the inlet, and made ready to go ashore in the zodiacs. The mist gradually dispersed and it became clear that we were in for a perfect day, like a fine September in England.

As the zodiacs were being launched, a herd of sea-lions (Hooker's) appeared and began diving and frisking round the ship in the calm water. We could see a great many more here and there along the rocky shores of the bay. However, we left them to themselves for the time being, running a mile or so back along the inlet and finally scrambling ashore on a stony beach under steep, vegetation-covered bluffs. Here we found any number of pretty birds on the wing – the small black-and-white shag (*Phalocrocorax campbelli*), little red-billed silver gulls, by far the daintiest and most charming seagulls I've ever seen, and light-grey antarctic terns hovering and dropping over the inshore water.

As we began an arduous clamber off the beach, we came upon an erect-crested penguin, standing solitary among the pebbles. These are small penguins, about eighteen inches tall, with two compact yellow plumes, like ears, each about as long as a human middle finger, or perhaps a little longer. However, this poor fellow couldn't afford us much enjoyment in coming upon a new species, for he was clearly sick. There was obviously more wrong with him than mere moulting. He was listless and hunched; and he was still there, motionless, when we returned about three hours later.

Once we had got up and off the beach, the actual climb to the open, higher slopes was not really hard, for there was not all that much height to be gained. However, it was made more difficult by the going underfoot – low scrub, everywhere ankle-deep and in places knee-deep. In the gullies the going was easier, but once out of a gully one was immediately faced with thick ground-growth and dense bushes. Lord knows what all the plants and ferns were: I tried to take note of them (apart from cursing them), but could not really hope to be comprehensive. There were a great many of the two-foot-tall, wild yellow lilies

(*Bulbonella*) common on the island; a little purple gentian; a pink saxifrage in boggy places; clumps of grey-leaved, small, chamomile-like *compositae*; and broad-leaved liverworts, bigger than any I had ever seen before.

At last we got on to the upper slope, below the ridge, where the ground became more turfy and open; and here the royal albatrosses were sitting on their nests.

Despite all that I had been told, I was unprepared for the sight of these birds. It fairly took my breath away. No words can describe their majesty, dignity and serenity. Imagine a wild, rocky hillside, broken with peat hags and rising to a windswept ridge. At intervals of two hundred yards or so are situated the nests – the usual circular 'castles' of earth and wispy grass. On each sits an unbelievably large bird, tranquil, unafraid, alert yet rapt in its own concern either of incubation or of guarding a chick from the skuas. Their beauty is like that of cliffs or caves. At close quarters the sheer size is startling and their statuesque, preoccupied detachment struck us all with a kind of awe. None showed either hostility or alarm, or made any move as we gathered round its nest (the males and females share spells on the nest and at foraging), but on the contrary gave the impression of being able magnanimously to spare us – whatever we might be – a little attention, provided we behaved properly – which it felt sure it could rely on us to do.

These albatrosses are fully as big as hen turkeys. Their weight must be about 16 or 17 lbs. The head, shoulders and breast are snowy white and thickly flocculent, the speckled wings charcoal grey. The great tubular bill, mushroom pink or pale mauve in colour, has a very noticeable, dark-green line along the lower edge of the upper mandible. The jet-black eye, alert and quick, has a nictitating eyelid.

From time to time the fluffy grey chick would look out from beneath the parent, but after a short spell usually seemed glad to get back into the warm. Above the nesting birds many more albatrosses were flying. These seldom beat their wings, but simply glided on their twelve-foot, outstretched spans. Whenever one passed above our heads we heard a gentle, here-and-gone *swish*, but nothing at all resembling, for example, the rapid *whoop, whoop* of swans' wings. The sound was simply that of a large, feathery body displacing the air. As often as a skua flew over – for even among these great birds the skuas are incessantly

Campbell Island. Royal albatross (Diomedea epomaphora) and chick. (Note: Chick menacing but not parent)

on watch for an unprotected chick – the sitting parent would raise its head skyward and clap its beak several times with a sharp, clacking noise of deterrence, loud enough to make nearby humans jump. The skua, which was no doubt used to it, merely flew on.

It amused me to see so many of us gathered about a nest, within a few feet of the bird, as though paying court. Tough Jim Snyder, Per the Swedish doctor, Big Dave and his pretty little Brazilian friend Juanita, Mr Thomas the gladhand, ocker Bob Leslie, Peter Hirst-Smith (still exhausted from lugging a 45-lb bag of cameras and lenses up the fell), Keith Shackleton and several more: and in the centre the unruffled, self-possessed bird, looking calmly from one to another as though aware of the admiration. In spite of all the cameras, I couldn't feel there was any vulgarity in the scene; for not a man but was deeply impressed, not a woman with any motive but to gaze in wonder.

Several of the albatrosses seemed actually to enjoy being gently stroked on the back. Others, however, would let it go on for a bit, then turn their heads and gently nip the stroking hand between their big beaks, as much as to say, 'Well, all right; but enough's enough.'

I've never seen the like. With the single exception of the herd of finback whales, it was the most memorable experience of the trip.

The breeding cycle of the royal albatross is so long that it mates and nests only once in two years. They mate in spring, round about September. Nest preparation takes 30 days and the incubation period is 78 days. After that, the chick is in the nest for about 236 days. This gives a total reproductive cycle of 344 days. (Yet the wandering albatross's cycle is even longer – 387 days.)

As usual in the tube-nosed family, the male arrives first at the breeding place, where it sets up home somewhere near (but not actually on) the same spot that it occupied two years earlier. There is a courtship period of about one month. Mating at Campbell Island takes place several times during October and November. Incubation is shared in long shifts of several days (even weeks), while one partner is away, fattening up after fasting on the nest. The chick is guarded day and night during its first five weeks, by which time it is large and tough enough to

*Parent 'clappering' to
deter skua*

defy a skua even when both parents are away at sea. Still, the
horrible skuas do sometimes succeed in snatching up an egg or
a young chick – more often, probably, from inexperienced
albatross couples breeding for the first time.

The chick grows steadily though slowly on the large but
infrequent meals brought by its parents at intervals of several
days. It has to endure the bitter sub-antarctic winter – snow and
gales – alone and unbrooded on the nest. However, it is well
covered with thick down and its first feathers, and it is comfortably dry in its hummock nursery, which it keeps clean, for it
squirts its waste matter clear of the pedestal. At last, when it is
about seven months old and after much preliminary exercising
of its wings, it takes off for the sea, having for some time been
neglected and unfed by its parents, and lost much of its thick
insulation of baby fat, which is replaced by the first, impervious
flight plumage.

During their next few years, adolescent royal and wandering albatrosses travel on the prevailing west winds of 40° to 60° South latitude right around the world, untroubled by any needs other than to use the strong winds for soaring, and to find their fish (chiefly squid) food, which they take by alighting on the sea. At three to six years old, they return to reconnoitre their birthplace and start displaying and seeking partners. However, the hen seldom lays her first egg until she is nine or ten years old.

If an egg or chick is lost early in the season, the parent birds abandon the empty nest, but return all the earlier next spring to make their next attempt. Following successful breeding, however, they are physically incapable of breeding again until they have had a 'sabbatical year' of rest and feeding at sea. Accordingly we found, on Campbell Island in the late antarctic summer, some pairs incubating, but in other nests full-grown young about to take off on their transoceanic wanderings.

Coming down, we encountered three birds together – two males competing for a female. The males were confronting one another, each raising his bill vertically to the sky and clacking vigorously. It was all display – a bit of pushing and no actual fighting. Real fighting, one would imagine, is beneath the dignity of royal albatrosses. The female watched, turning her head this way and that. The business seemed likely to take some time. We passed on, and I didn't see who won.

Here I digress to tell a strange story which I heard while we were at Campbell Island. In a remote place near the head of the inlet may be seen the ruins of a cottage, together with an unmarked, untended grave. Legend has it that the grave is that of an illegitimate daughter of Prince Charles Edward – 'Bonnie Prince Charlie' – who was compelled to live in this desolate place (as, later, Napoleon on St Helena) so that she could not become a focus for a further Stuart rebellion. Captain Ross records meeting a very old Scotch lady who lived on the island, and adds that on an earlier visit he saw heather near the ruins, which someone must have brought and planted, since *Ericaceae* are not indigenous to the Antipodes.

The legend is plainly rubbish. I have since put it to Sir Peter Scott, who is equally incredulous. Prince Charles Edward died in Rome in 1788, aged sixty-seven, not having been within any territory of the English crown since 1746. His only acknowledged daughter, Charlotte, died in 1789. How could any other daughter of his – even assuming there was one – have been forced to come to a place like this, and by whom? Ross did not voyage in these waters until the early eighteen-forties. Gibbon Wakefield did not colonize New Zealand until 1828. It just doesn't add up. Whoever she may have been, the Scotch lady on the island can't possibly have been Bonnie Prince Charlie's daughter.

During the afternoon of this wonderful day we made a second expedition, to visit the largest of the sea-lion colonies along the shore. There were perhaps thirty sea-lions altogether – bulls, cows and pups – either along the beach or lying among the grass and bushes above it. They were not at all afraid of us; on the contrary, we had to be a little wary of them. Whenever anyone came too close, they threatened him by a display of gaping, the bulls uttering their roaring cries and lurching a few feet towards the intruder. I came to the conclusion, however, that in fact they were less minatory than inquisitive – like cows in a field. They wanted to see what we were. If they found you near them they would come closer, just as cattle do, but if you stood your ground, all they did was to stop a few feet away, poking their heads forward and sniffing curiously.

These sea-lions were very beautiful, the adults having thick, smooth, dark fur, while the young were light tawny – a few even creamy – in colour. They had endearing faces, gazing gently and sleepily at us from huge, brown, liquid eyes. It was delightful to watch the parents playing with their pups, some of which were still mere babies, only about three feet long and looking like very large, furry kittens. We watched a little, fawn-coloured female pup nipping and tussling as it played at attacking its bull parent, which could have flattened it like a steam-roller. Another parent, a cow, as it groomed its pup, was fondling and caressing it with obvious affection. (In the Canadian harp-seal slaughter, about 150,000 such pups are battered to death each year before their dams' eyes. The official Canadian line is that the dams don't really mind.)

The dignity of these graceful, playful animals was quite different from that of the royal albatrosses. The former had an appeal rather like that of fine horses. One respected and admired the albatrosses. One felt affection for the sea-lions.

10

ENDERBY AND SNARES

Now the weather became startlingly warm and sunny. We'd thought it pleasant enough at Campbell Island, but next day, when we reached Enderby Island in the Aucklands group, it was like a fine English June. After the deep Antarctic and the ice, it was strange to feel no wind at all and the sun warm on your skin. We had visited the bottom of the monstrous world and were now coming up the opposite side.

Enderby Island, which is uninhabited, is a New Zealand nature reserve; low-lying, partly wooded with low trees and bush, partly open, sandy dunes - golf-course country, as some wag remarked. On this beautiful morning of early March it appeared most *luxe, calme et volupté*, and the zodiacs were filled with no hesitation at all. 'Mawnin', Ri-i-ich'd,' said Carolahn Best happily, meeting me at the sally-port. We had warmed to each other since that far-off, 4 a.m. meeting in Buenos Aires when she first introduced herself, and I had come to enjoy her Texan drawl and her buoyant Texan approach to everything in general. Wander through the streets and note the qualities of people. In fact, during four weeks on board, we eighty passengers had become a ship's company with a lot of *esprit de corps* and mutual esteem. Perhaps this was partly because we were mostly so elderly - older people fall out among themselves much less than the young - but I like to think that it was also due to the wonderful experiences we had shared. Mr and Mrs Frederiksen - now become Rod and Helena - Bob Leslie, Carolahn, Nellie Hartwell, Tony Irwin, old Betty Robinson and the rest - we had all been among the finback whales, we had stood together in Captain Scott's hut, we had felt the terrible wind at −56°F, walked among the king penguins, seen the royal albatrosses. We weren't those greenhorns who had joined up at Buenos Aires. We thought well of each other.

As we approached the shore of Enderby the soft, bright air was full of bird-song, melodious, clear and full, recalling that of blackbirds or mocking-birds. It was only now that I realized that we had heard none for weeks – hardly a cackle or a squawk, for albatrosses are silent and southern gulls (as I've already remarked) don't squabble and yell like European herring-gulls. The beautiful, placid sound added still more to our sense of relaxation and return. What we were hearing was, in fact, the song of the New Zealand bellbird.

Part of the coastline of Enderby consists of a length of low cliff of strange appearance; a close-set zig-zag, perfectly regular, looking as though a giant had pleated a great length of thick, grey paper, then pulled it partly open and set it upright. This is columnar basalt. Each column is no more than four or five feet across and perhaps forty feet high. The regularity was very striking. One column had fallen out of the long row and lay vertically upside-down, its one-time head resting on the shore below. Here and there along the cliff were deep caves, and into these our zodiacs went. Inside, the sea was fairly turbulent. The blind wave feeling round his long sea-hall. At the helm of an outboard motor, however, any one of Big Dave's lads were perfectly capable of holding a rubberized zodiac steady within a few feet of a rocky wall with swell surging against it. I was amused by the way in which they communicated with each other and the ship by means of pocket radio sets. 'Jim Snyder, Jim Snyder, Jim Snyder, do you read me? We have a yellow-eyed penguin in this cave, about five hundred yards east of where you are. Over.' If this was technology, it was harming nobody and spoiling nothing – to say the least.

The yellow-eyed penguin is about the same size as a gentoo, with a grey head and – oddly enough – pale yellow eyes. It does not frequent the colder southern latitudes, but inhabits the seas and islands of southern New Zealand and its neighbouring, sub-antarctic islands. The greyish-pink bill is tipped brown and the back is greyish charcoal. For a penguin it is unusually timid and tends to avoid humans, nesting in cavities or in dense bush (woodland). In moult, however, it is often torpid and apathetic and can then be approached more easily.

Landing on the grassy, open shore, we dispersed and set off, in groups or alone, whither we would. There was an atmosphere of fantasy and euphoria, induced by the strangeness of the place

Enderby Island. The 'pleated' coastline

Yellow-eyed penguin (Megadyptes antipodes)

South Island 'tomtit'
(Petroica
macrocephala) *in
dense woodland*

and the extraordinary weather. I wasn't the only person to be reminded of *The Tempest* or *Dear Brutus*; a company of people scattered and wandering in an unknown but unhostile and exciting environment.

I wandered off alone into the thick, low woodland along the top of the basalt cliffs. Here I came upon passerine birds which approached through the branches and came almost to the hand – out of sheer curiosity, or so it seemed. The bellbirds were singing a smooth, four- or five-note phrase, very rich and sweet – sunshine turned into sound. They are not much to look at – something smaller than a blackbird but bigger than a chaffinch, shaded olive and brown, with a touch of white on each shoulder. Another local inhabitant was the little, black-and-white ball of a bird which New Zealanders call a tomtit. (It is not a tomtit at all, but a species of flycatcher, as evidenced by the pointed, sharply projecting little beak.) These birds came hopping about, only a foot or two away, in the dim light and the low branches, until you might have thought some spell had deprived them of natural fear.

Here on Enderby there were flowers at last; real blooms, not merely the cold-pinched little chickweeds and crowfoots of Macquarie. Growing profusely was a flowering myrtle, *Metrasideros rata*, some ten feet tall, with light-green sprays of oval leaves and a red-plumed bloom rather like a modest version of the bottle-brush. On the open cliff-top was a tiny daisy, unknown to me but undoubtedly a *Bellis* of some kind. Another pleasant find was a deep-pink pimpernel with paired leaves nearly orbicular, very like our own bog pimpernel, the *Anagallis tenella*.

On this island, too, there were a great many sea-lions, the pups very roly-poly and frisky in the warm sunshine. It was rather startling to come upon – indeed, almost to tread upon – mothers and pups lying secluded in the thick woodland, the mother suddenly rearing up among the fern only a few feet away from the intruder, to see what it was that was coming. However, these were not by a long chalk the oddest inhabitants. No; the strangest animals on Enderby are the *rabbits*. These (like the wekas) were originally introduced by the old-time whalers for food, but they have long been left to their own devices – though from time to time efforts are made to keep them down, for they have destroyed a lot of vegetation and large areas of the open turf are unstable, being honeycombed with their warrens. From what cause I do not know – in-breeding, perhaps – these huge rabbits ('Look,' said someone, 'there's Bigwig! He must have emigrated') have extraordinary colours; not only black, white, and black-and-white, but also ginger and, believe it or not, a kind of blue. And I don't mean the Persian grey which is sometimes called 'blue' (e.g. in cats) but an actual *indigo*. During the afternoon, as Ronald and I were sitting in the woods, through the undergrowth came two ginger-and-indigo rabbits, which foraged unconcernedly near us for some little time.

On parts of the open grassland, which stretched away smooth as a great lawn (there are grazing herds of wild cattle on the island, though I came upon none), flocks of brilliant, emerald-green parakeets (*Cynoramphus*) were feeding. Their plumage – scarlet heads, but all the rest green – was so vivid that it glowed and spun before the eyes. It was impossible not to feel excited ('Oh, look!') when they flew up all together in a flock, forty or fifty at a time. There were dark-plumaged pipits, too, very like the rock pipits of the Isle of Man, and Ronald saw a New Zealand falcon, though I missed it.

Hooker's sea-lions on the beach

Jim Snyder and friend on Enderby

However, undoubtedly the most remarkable bird on Enderby is the flightless Auckland teal (*Anas aucklandica*). This is a very rare bird indeed, being found only in the Auckland islands. It lives along the rocky shore at the foot of the low cliffs, and forages among the thick beds of sea-kelp and *Duvillea* (on which it can walk, of course). Here it survives because there are almost no predators. It is only about eight inches long, in colour a kind of dark cinnamon or mottled red-brown, with vestigial wings. Not much to look at, really, but nevertheless one of the most uncommon birds in the world. It was not easy to seek out, but one of our company, the zodiac-tugging Jim Snyder, had the luck to come upon and photograph it.

Along the ledges were colonies of black-and-white shags. These have a scarlet patch under the chin, very conspicuous when the birds are seen from below. They, too, were not concerned to avoid us.

We spent all day on the island and came back to the ship quite tired out. The fine weather continued and the next morning we found ourselves off Snares Island, an almost inaccessible fang of precipitous rock a hundred miles south of New Zealand. This, too, is a nature reserve and landing is not permitted, but we took to the zodiacs for a run round the inshore water.

Here was the greatest possible contrast to the soft, green banks of low-lying Enderby. As we came in under them, the grey cliffs rose sheer above us, bleak and rugged in the wild solitude. Among the offshore swell were fairy prions – very attractive little birds – and diving petrels (which can hardly fly) were fluttering about, emerging unexpectedly from the waves. Along the lower ledges of the island fur seals were lying, occasionally lumbering down into the sea to forage.

Snares Island is one of the few nesting sites of Buller's albatross, exclusively a New Zealand species. This, the smallest of all the albatrosses, with an average wingspan of a mere six feet, is also one of the most beautiful; the head and neck a smooth grey, the big tubular bill vividly striped yellow and black. The entire island, every cliff and ledge, was covered with Buller's nests. We could see sitting birds everywhere. Peter, looking up at the cliff above him, saw a young chick snatched by a skua

The cliffs of Snares Island (zodiac on right)
(pages 144-5)

while the parent bird was momentarily off the nest. The air above the high summit of the island was full of these albatrosses, gliding gracefully round and over the inaccessible precipices. We felt very small, tossing up and down at the base of the cliffs; each boat, with its sixteen passengers, a mere speckle, like a grain of pepper on a table-cloth.

During this visit of ours the Snares or Buller's albatrosses (Dio-medea bulleri), strangely, upon the approach of winter, were just beginning to incubate their single eggs in cup nests plastered against the cliff ledges. This apparently untimely nesting season in such a stormy latitude is a curious divergence from the régime of Buller's albatross farther north, in the warmer Chatham Islands, where a slightly smaller, darker race follows a more normal cycle of incubating the egg in the spring (October–November) so that the young birds fly in the autumn (April–May). The Snares albatross launches on its first flight in late winter. The reason for this is not clear, but as Nature has a reason for all animal behaviour, the secret may one day be discovered.

The professional ornithologists, like Ronald, were delighted to find among the caves one or two of the rare Snares penguins, a small species local to these seas and islands. For my part I was so much affected by the lonely, gaunt starkness and towering height of the place, with its tens of thousands of albatrosses, that I hardly had any attention for individual things. This was a place which made a single, overwhelming impression. It had been like this long, long before there were men on earth, and it would be unchanged in a million years. No doubt that was one reason why we felt so small. It wasn't merely a matter of the size of the cliffs. As Richard Jefferies remarked, 'You cannot tell what century it is from the sea.' Or from Snares Island, either.

11

STEWART ISLAND: OUR JOURNEY'S END

Finally, after more than a month at sea, our ship reached Stewart Island, about twenty miles south of the southern tip of South Island, New Zealand, in latitude 46°–47°. This largely wild island – archipelagic, estuarine, afforested, ferny, humid – is about forty miles long and twenty-five across. It is divided in the middle by Paterson Inlet, a broad, irregular estuary some eleven miles long and three miles across, full of bays and islands. The biggest of these islands, Ulva, a mile and a quarter long, is a nature reserve, as also is the smaller Native Island which, together with the peninsula they call The Neck, closes the inlet at its eastern, seaward end. Just north of these, at the eastern end of the northern shore, lies Halfmoon Bay (also called Oban), the island's only township (population approximately 450) which is very like an antipodean equivalent of Dylan Thomas's village in *Under Milk Wood*.

Most of the island is low-lying, but it has two low ranges of hills, one in the north and one in the south. Its highest peak is Mount Anglem, in the north, 3,214 feet. A good three-fifths of the whole place – the northern area and the larger, southern area – comprise two nature reserves of wild, dense bush, watched and maintained by rangers working for the New Zealand government. A lot of this forest is accessible only by helicopter. Here, if anywhere (for the bird has become very rare indeed), is to be found the kakapo (*Strigops habroptilus*), the green-speckled, flightless parrot, rather bigger than a chicken, which was once fairly common throughout New Zealand. However, it is not only scarce but nocturnal and shy. Reluctantly, Ronald and I decided that we could not practicably form any plans to try to go and see it.

Many place-names on the island – as all over New Zealand – are Maori: Kopeka Island, Pikaroro Point, Ringaringa beach, Taukihepa Island and so on. Other names sound like a mixture of the Californian gold-rush and Robert Louis Stevenson: Port Adventure, Murray's Mistake, Chew Tobacco Bay, Dead Man's Bay, Long Harry Bay, Pearl Island, Little Hellfire beach.

Landing, we found ourselves at once in the sort of easy-going, amateur atmosphere characteristic of remote communities; and although we left our gallant ship with regret, at the same time we felt excited at the prospect of a week among rangers and watermen, tree-fern forests and bird-haunted islands.

In Maori folk-lore the North Island of New Zealand was originally a great flat-fish hooked and pulled up by Maui, the trickster folk-hero. The South Island was Maui's canoe, while Stewart Island was the anchor of the canoe – Te Punga o te Waka a Maui. All the same, there is no evidence of Maori occupation of the island earlier than the thirteenth century. Captain Cook, the first European to reach it (in 1770), thought that it formed part of the mainland and accordingly named it 'Cape South'. Nearly forty years later, in 1809, William Stewart, first officer of the *Pegasus*, charted parts of its coast, established that it was an island and gave it its European name. It quickly became a sealing and whaling base, and by 1830 there was a resident population of sealers, with Maori wives.

Five or six miles out from Halfmoon Bay lie the Muttonbird Islands (with appropriate names: Women's Island, Jacky Lee Island, Te Marama, Kanetetoe, Maria Higgins Rock, etc.). The so-called Muttonbird is none other than the highly numerous sooty shearwater (*Puffinus griseus*). At least, this is the kind most commonly taken, though all species of petrel are taken on occasion, and are good to eat. Taking young muttonbirds is a somewhat barbarous business, dating, among the Maori, from time out of mind – long before the coming of the Europeans. Shearwaters nest in sandy burrows. Those breeding on the Muttonbird Islands lay their eggs in November or thereabouts and hatching takes place round about the end of December. About three months later, in early April, the parent birds desert the young and migrate north. Left to itself, the deserted chick duly matures and follows independently. But before that, the muttonbirders are at work. The traditional method of dealing with a young muttonbird is to pull it out of the burrow, wring its neck and then, twisting the head round, eviscerate it with its own beak. The birds were (and are) either eaten fresh, or salted down. Nowadays many are deep-frozen for the commercial market.

Muttonbirding is a jealously guarded, inherited privilege and under New Zealand law only descendants of the original Maori owners of the Islands are allowed to carry it on. Consequently, the only 'European' muttonbirders are husbands or wives of Maoris, and their children. Muttonbirding is unique in being the only traditional Maori food-producing activity to have remained unaffected by 'European' competition. The limitation by law is, of course, all-important, for the muttonbirders avoid – or at all events are supposed to avoid – over-exploitation of the birds.

In present-day New Zealand, muttonbirding faces a certain amount of opposition. Its opponents feel that while it may have been all very well for the old-time Maoris (a lot of whom were very hard up for protein, to an extent which often drove them to cannibalism) to take muttonbirds for their own consumption, to-day too many are taken, and the thing has become an unjustifiable wildlife slaughter for commercial gain. Though I'm not a vegetarian, I confess that I personally find it distasteful that large numbers of young, defenceless, wild birds should be pulled out of the nest and put to death solely for commercial gain.

In small, remote communities nothing is ever hurried. Good manners demand that things should be done leisurely, with plenty of time for talk and deliberation. We drank beer and played pool with the fishermen in the local bar (which was convenient, as it formed part of the only hotel, where we, of course, were staying). We met the wildlife rangers, Jack Hinton and Dickon James, and I listened fascinated as they, Ronald and Eric Hosking talked about famous ornithologists they had known, and about the problems facing New Zealand ecology and wildlife today. One of the difficulties of rangers in Stewart Island is to control and keep down the deer, which do a great deal of damage, often eating down natural vegetation in the protected wildlife areas almost to the point of extinction. In this they have the assistance of both rats and 'possums.

Dickon, I am sure, must be the only man to have 'bulldogged' deer by helicopter – and even he did it once too often, as he himself described to me. 'Bulldogging' is running down a quarry (e.g. deer or cattle) and taking it alive, often by actually leaping on to it and thus bringing it to a halt.

'These deer,' said Dickon, 'are white-tails, with a live weight of about two hundred pounds. They do an awful lot of damage, but of course we hate killing them and try to avoid it as much as we can. We prefer to take them alive and send them to the children's zoo in Hamilton. But of course you can't really follow them through the bush on horseback, so I had the notion that we might be able to follow them by helicopter and jump on them from the door. Anyway, I reckoned it'd be a bit of sport to try it. I got two that way, actually. The first one we followed over the scrub and overtook it in a fairly clear patch. Then I jumped on its rump and grabbed it by the ear. It winded me, but I was able to bring it down. Then I tied its legs, blindfolded it and loaded it into the chopper. The other much the same. Still, I admit I tried it once too often. The third one, we were doing about thirty miles an hour and we were just behind it, so I jumped for it, but unfortunately at that moment it dodged round a tree and I missed it and hit the tree. I thought I'd broken my collar-bone, and I had a bruise all across my chest for about a fortnight. Still, it was a bit of a lark,' concluded that amazing man.

One morning, Ronald and I had been up in the hills above Halfmoon Bay, watching any birds we could find, but principally the New Zealand native pigeon (*Hemiphaga novaeseelandiae*). This very beautiful bird, a good deal larger than the English wood-pigeon (perhaps twenty inches in length) is white-breasted, with a slim, graceful head and neck, the back and wings olive, varied with deep, iridescent purple. It has a most striking, deep-red eye and a dark tail with a fawn-coloured border. The head and upper breast are emerald green. The hooked bill is orange and the legs are covered with a white flocculence down to the toes (reminding me of a Kathak dancing-girl's trousers). We were lucky enough to be more or less concealed when three or four of these splendid birds alighted in the trees a few yards away, so we were able to watch them for some time. In fact, it was all so absorbing that we quite forgot that Dickon had promised to meet us in the early afternoon and take us out in his jeep to another part of the bush.

'You know,' said Ronald, looking at his watch as we sweated along at a brisk walk, making the best of our way back through the humid noonday, 'we're going to be late: we can't avoid it. But I tell you what; the first house we come to, we'll drop in and ask whether we can use their telephone to let him know.'

'Good idea,' said I. 'It's a friendly place. Whoever they are, they won't mind.'

In due course we came to a nice-looking, solitary house on the edge of the hills, overlooking Halfmoon Bay below.

'Come on,' said Ronald. 'This is it. Look, they've got a 'phone all right. There are the wires.'

We walked up the short drive and rang the bell. No answer. We knocked and banged. Still no answer. The front door, however, was slightly ajar. Ronald pushed it open and we ventured in and called out. Not a sound. Clearly the house was empty.

'Never mind,' said Ronald. 'There's the 'phone, look, over there. 'Won't take a moment.'

He picked it up and the operator asked him for his number. 'I'm afraid I don't know the number,' said Ronald politely. 'But will you please put me through to Dickon James, the ranger?'

'Yes, I will,' replied the operator genially, 'but first of all, d'you mind telling me who you are and how you come to be in my house?'

Later that week we met Nancy Schofield, the operator, at the flower show, and told her that it was us. Wonderful place, Halfmoon Bay.

One evening, when we were having a drink at Dickon's house, he told us that he had a fairy prion in a box down at his wildlife office. He had picked it up exhausted after it had been blown ashore in storm and high wind a day or two before, but now that the weather had turned he was going to release it that night. We accompanied him down to the office and he gave me the prion to hold and have a look at before we released it. It was the most dainty and graceful little bird imaginable, very neat and compact, as big as a nuthatch perhaps, its plumage banded white and smoky blue. It seemed quite happy to sit in my cupped hands. When Dickon took it outside and tossed it upward it immediately took wing into the gathering darkness.

The most striking birds to be seen along the foreshore were the all-black oystercatchers. This species, whose plumage is entirely black all over, is solitary; you never come upon more than a pair (or a pair with chicks) in any one place. They breed on the shore and stick to one spot all the year round. They are rather larger than the common, pied oystercatcher, with groups of which they will consort (as black-backed gulls do with herring gulls). They have pink legs and very conspicuous orange eyes (matching the long beak), which show up well against the black plumage.

As well as the rangers, another great local personality was Bill Redpath, at eighty-seven the grand old man of the island and a grandson of the Rev. Wohlers, a German missionary who spent over forty years on and around the island between 1844 and 1885. (Wohlers's grave, with an unexpectedly handsome monument and a well-phrased, moving inscription, stands on a high point above Ringaringa Bay.) Bill was an old acquaintance of Ronald's, and we were able to go on expeditions with him in his launch. Amongst other things he told me was how his right shoulder had been smashed at the first battle of the Somme in July 1916. 'It was very bad,' he said. 'But these days I can't remember things so far back, you know. Why don't you ask me about the kakas? I can tell you all about them.'

The kaka (*Nestor meridianalis*, not to be confused with the kakapo, *Strigops habroptilus*) is a big bird of the parrot family which inhabits thick bush; and some of these, under the guidance of Mr Redpath, we were lucky enough to come upon one afternoon, after we had waded ashore from the launch and climbed up into the dense woodland of a bluff standing above the tidal shallows. We were admiring an eighty-foot tall red pine (*Dacrydium cupressinum*) with a trunk sixteen feet in circumference at four feet from the ground. (Bill reckoned it must be five hundred years old.) It was surrounded by a green mound a foot or two high, composed of its own fallen bark. Suddenly we heard nearby the call of the kaka – a loud cry on two notes, followed by a croaking screech. (In fact the kaka has a variety of cries, but this is the commonest.) We kept quiet and after a little two kakas duly appeared. They were a soft brown all over, with bronze-coloured faces. Under the wings, however, kakas are a deep crimson, so that when they open their wings to fly the effect is startling and very beautiful. They have massive beaks, with which they slash the trees for sap; and in fact they can bark-strip and kill a tree in this way. They were not easy to keep under observation in the thick woodland, but in the course of the next hour we managed to get several quite good sightings as they came and went about their business.

In the course of the week we made several launch trips, either with the rangers or with Mr Redpath, or both; and in this way we were able to finish our long travels with some splendid ornithological (and other) finds up and down the windings of Paterson Inlet. We saw our final species of albatross – the Shy albatross – close to, for it came to eat pieces of fish which we threw overboard to attract it. The Shy albatross – like Buller's, which in fact was with it in the inlet – doesn't go far south. It has a wingspan of no more than about seven feet. It looks as though its head were enclosed in a smooth, white helmet, with dark eye-slits – a close-fitting mask. Indeed, it looks, at close quarters, rather like a soldier in a science-fiction story; distinctly grim. Buller's albatross, seen close to (as we had not been able to see it at Snares Island), was grey-headed, rather smaller than the Shy: the black-and-yellow beak very striking.

That same afternoon we saw our last two species of penguin. The first was the tiny blue penguin (*Eudyptula minor*), the smallest species, which 'would lose a fight with a duck', as Dickon remarked. It was diving and 'porpoising' near the launch, but it was so small that for some time I couldn't get much of a sight of it. In colour it is not really blue, but grey (Persian blue), with a fluffy appearance. It is no more than a foot long. A very attractive little bird.

In the bush on an island Dickon stalked and caught a yellow-eyed penguin. As already mentioned, this species tends to flee from man and prefers to conceal itself, so I was lucky to have this chance to see it at close quarters. The yellow eye is a remarkable sight – a smooth, dull yellow, the pupil hardly visible, so that the bird looks as though it had wall eyes.

Coming back, we happened upon a fur seal killing an octopus before eating it. My companions reckoned that it must have gone to a depth of about ninety feet to catch it. Albatrosses will eat fish alive, but seals will not, and this seal was making a great business of killing the wretched octopus, beating it from side to side in the water and then throwing it up into the air. Gulls hung around for the fragments, and the grisly business went on for well over five minutes. Finally the seal finished off and ate the octopus and dived (or 'hyper-ventilated', as they say – dived holding its breath in a big way), presumably, for another; but we saw it no more.

That week in the green, estuarine humidity of the inlet (it seemed positively unreal to have an air temperature of 14°C) was enough to make us wish we had longer to spend there. We

The fur seal killing the octopus

used to set out from a sheltered anchorage near the three little islands known as Faith, Hope and Charity (appropriately, Charity's the biggest) and make for Ulva Island, for the mouth of the Freshwater River or even, daring the choppy waves, for the open sea beyond The Neck peninsula and so south to East Cape and Port Adventure. (Port Adventure is just a cove with a solitary campers' hut among the trees.) I remember the white-fronted terns (bigger than the arctic tern and unique to New Zealand) fluttering and pouncing in the calm water; as recognizably and unmistakably terns in their behaviour as the black terns we had seen in the sub-antarctic: the spotted shags and blue shags – a dimorphic species, some black and white, some black with red chins: a flock of more than two hundred godwits, forming wide crescents in flight as they made their way across the Inlet on the mudflats: and a great flotilla of hundreds of black Australian swans, covering acres of the water on which we drifted silently down with the wind, our engine stopped and we ourselves half-hidden behind the gunwales, in hopes not to alarm them. Nevertheless, we didn't succeed in getting all that close, for these swans are shy.

I shan't forget the weka I met face to face, walking towards me along a narrow trail through the bush on Ulva Island. Intent on wherever it was going, it simply walked past me, just as a human being might. I felt like murmuring some casual word in passing – 'Nice day, weka.'

On Ulva, too, I saw the big stick insect *Diapheromera*, bright green and a good five inches long, lying in wait for prey, just like the mantises I remember in Palestine. There must be a lot of seepage on that island, for everywhere we saw huge liverworts, as big as the palm of your hand; and the so-called umbrella mosses, about two inches tall, like little green parasols, perfectly shaped for moisture retention. Here, too, we saw, carpeting the ground, a tiny wild lobelia (*Pratia*, I believe, is its name) which under a lens (it's so minute) revealed that its five white petals are purple-streaked and yellow at the centre. Its local name is the Banker's Flower, because it's open from nine till three. This reminded me of our own English Goat's Beard, also called Jack-Go-To-Bed-At-Noon, because that's when it closes.

I wrote my valediction to Stewart Island in an entirely correct and traditional way. There is a tree, known as the muttonbird tree, with thick, soft leaves, the larger ones as big as nine square inches, glabrous and white on the undersides. Until quite recently, Stewart Islanders used to use these leaves instead of postcards, writing short communications on them and then addressing, stamping and posting them. You can still post them and people do (they take ballpoint very well), but nowadays they have to be put in envelopes. It pleases me that I have sent a letter of thanks and farewell on muttonbird leaves.

So here our journey, which had taken us under the world, from Cape Horn to New Zealand, ended at last, and what our shipmates had called the Gang of Three dispersed; Peter back to his photographic work in London; Ronald to return home to Auckland by easy stages, travelling up the marvellous fjord country of South Island in company with Eric and Dorothy

Hosking; and I to rejoin the ship, as she returned to Halfmoon Bay from Lyttleton, to set out on yet another voyage; this time a warmer one, up the Great Barrier Reef and through the Coral Sea to Papua New Guinea and its remote, tropical archipelagos. That is another (and no less wonderful) story, and one day I may try to write it.

Thank goodness I have – at least in some small way – seen the Antarctic, and that, too, in the company of a great naturalist like Ronald Lockley. To me it is reassuring and comforting to re-member those vast, bitter solitudes, the most richly populated, fascinating and rewarding wildlife area in the world – the innu-merable penguins in their rookeries, the sheathbills, the huge elephant seals, the predatory skuas and above all the incompar-able albatrosses for ever gliding round the world on their out-stretched wings. It'll take a bit of ruining, will that. Yet even this inaccessible, icy Eden is precarious. Some of you who read this book may well live to read also of the destruction by human greed of the Antarctic, as even now you are learning of the felling for commercial gain of the equatorial rain forests. (And God only knows what the final effect of that will be.) Already, during the past hundred and sixty years, many species of whale have been hunted and slaughtered to the point of extinction, and all the efforts of such great conservationists as Sir Peter Scott have been able, as yet, to achieve little more than a ray of hope for the future. Is Joseph Hatch securely in his grave? Plainly not, for as I write these lines, the Japanese, in co-operation with the Argen-tinian government, are proceeding with plans to slaughter ant-arctic penguins in order to extract the oil from their bodies for commercial profit.

Those finbacks among which we sailed with such delight – how many of them are still alive now? Those sinister plans – on the part of Russia and certain other countries – to 'harvest' the antarctic krill: if they go forward on any scale, what will be the effect on antarctic ecology and on the millions of birds and animals entirely dependent on this food-source? Whose Antarc-tic is it – ours or theirs?

Many of the wildlife areas of the world have been destroyed, or else diminished to mere fragments of what they once were. But it is still not too late for an international agreement to protect and preserve the Antarctic for its rightful inhabitants, which are among the noblest and most beautiful creatures in the world.

THE PHOTOGRAPHER IN THE ANTARCTIC

Peter Hirst-Smith

Going to the Antarctic was for me the fulfilment of one of those mysterious plans made for us in life: as a child one of my favourite books was *The Great White South*, full of pictures of ships stuck in the ice, brave heroes bound for the Pole and beautiful snowscapes, all photographed by the legendary Herbert Ponting; and later, one of the visiting lecturers to our school was a gentleman named Eric Hosking, whose magnificent photographs of birds and animals probably planted the seed which is still working away in my system. I could have had little idea then that one day I would be standing in Ponting's antarctic darkroom, or that I would be working side by side with Eric Hosking. Yet this is exactly what happened during this trip.

On setting out into what was quite unknown territory, my choice of equipment was based on two considerations: versatility and portability. After much thought I decided to use my 35 mm system, because the quality of the lenses is high and it is much less bulky than any other format. I took three camera bodies (one with an automatic meter, a flash unit and motor-drive), and a wide range of lenses: 20 mm, 28 mm, 35 mm, 55 mm macro, 105 mm macro, 135 mm, 200 mm and 300 mm. In addition to these I took two hand-held light meters, a number of filters and a good supply of power packs for the flash unit and motor-drive. I should have added a small tripod, which in the event I had to borrow. One of the hazards I foresaw before departure was the possible freezing of the lubricating oil at temperatures below −20°C. Our visit was in late summer, so this problem was minimized, but we still encountered weather cold enough to necessitate keeping the camera warm between one's body and clothes. At times lenses can mist over, too, so a good supply of Selvyt cloths is useful. There were, of course, things I didn't foresee: the motor-drive proved its worth in photographing birds in flight, but a spot-meter would have been valuable for calculating the exposure required for taking wildlife against brilliant backgrounds. Keeping salt water out of the equipment proved a major problem – three very expensive cameras (not my own, fortunately) were ruined during the trip because of their inability to swim! I was glad I had a really good camera bag, but I wish I had followed Eric's advice to take a good supply of plastic bags – you can now get ones that seal completely and will float. Alternatively you could take one of the totally submersible cameras made for photography at sea, such as the Nikonos. On a future occasion I would take two zoom lenses, a 35 mm to 80 mm and an 80 mm to 200 mm, to avoid having to change lenses so often. In some situations, too, I could have used one or two small flash-units, fired by a slave unit, to light one area of picture but not to take over as the main light source.

On most expeditions I would normally take the appropriate Kodak professional filmstock, but since we were to pass from temperate England through the tropics to the Antarctic and back again, colour shifts resulting from temperature variation would have caused problems. Amateur filmstocks have built-in inhibitors to compensate for this, so in the end I took a film which I knew would give excellent quality and colour saturation – Kodachrome 64. I took three hundred rolls of this, which turned out to be more than enough, plus fifty rolls of 400 ASA Ektachrome, and plenty of black-and-white as well. To cover every eventuality I also took some tungsten-type film – just in case.

The most serious photographic problem of all is exposure. Automatic through-the-lens light meters read snow as grey and therefore under-expose the white. The best solution is to make plenty of exposures, bracketing up to two stops to compensate.

The technical difficulties associated with photographing birds and animals in the Antarctic are of course unusual, but there is one great bonus: man's presence here does not naturally signify a threat to wildlife, as it does in so many other parts of the world, and it is possible to get within touching distance of such magnificent birds as the royal albatross with her chick, she worrying far less about you than about the predatory skuas overhead. On one occasion, when Jim Snyder and I were exploring the undergrowth on Enderby Island, we came across a female Hooker seal with her pup; far from being alarmed by us, she looked on placidly while the pup, honking with delight, came up to nose our faces.

Experiences like these, and the sheer beauty of the whole area, were overwhelming – the more so as all I could do was to photograph it. I hope something of what we saw and felt comes through to the reader.